MATERIALS AND APPARATUS IN QUANTUM RADIOPHYSICS

MATERIALY I USTROISTVA KVANTOVOI RADIOFIZIKI

МАТЕРИАЛЫ И УСТРОЙСТВА КВАНТОВОЙ РАДИОФИЗИКИ

The Lebedev Physics Institute Series

Editors: Academicians D. V. Skobel'tsyn and N. G. Basov

P. N. Lebedev Physics Institute, Academy of Sciences of the USSR

Recent Volumes in this Series

Proceedings (Trudy) of the P. N. Lebedev Physics Institute

Volume 98

Materials and Apparatus
in Quantum Radiophysics

Edited by
N. G. Basov

P. N. Lebedev Physics Institute
Academy of Sciences of the USSR
Moscow, USSR

Translated from Russian by
A. Mason

Springer Science+Business Media, LLC

Library of Congress Cataloging in Publication Data

Main entry under title:

Materials and apparatus in quantum radiophysics.

(Proceedings (Trudy) of the P. N. Lebedev Physics Institute; v. 98)
Includes bibliographies and indexes.
1. Quantum electronics. I. Basov, Nikolai Gennadievich, 1922–   II. Series:
Akademiia nauk SSSR. Fizicheskii institut. Proceedings; v. 98.
QC1.A4114 vol. 98   [QC688] 530s [537.5]   80-21931

ISBN 978-1-4757-5790-3   ISBN 978-1-4757-5788-0 (eBook)
DOI 10.1007/978-1-4757-5788-0

The original Russian text was published by Nauka Press in Moscow
in 1977 for the Academy of Sciences of the USSR as Volume 98 of the
Proceedings of the P. N. Lebedev Physics Institute. This Translation
is published under an agreement with VAAP, the Copyright Agency of the USSR.

PREFACE

In this volume the properties of free-running and pulsed lasing are considered for a ruby laser at temperatures near 100°K caused by increased induced ruby emission at lower temperatures. Various free-running lasing regimes are considered together with threshold conditions and spectral dynamics.

Processes of formation of local environments of rare-earth ion additives are studied in laser crystal of garnet structure. Various types of activator TR^{3+} ions are classified and analyzed in mixed garnets of various composition and structure, and interaction among TR^{3+} ions is studied.

Second-harmonic generation of laser light in powders and single crystals of organic substances is studied. The nonlinear optical properties of organic compounds are correlated with the crystal and molecular structure.

The applicability of the WKB method (of geometric optics) is justified in calculating the modes of optical resonators containing an inhomogeneous active dielectric medium, and the mode spectra, losses, and fields are calculated for resonators encountered in laser applications.

This volume is addressed to specialists in radiophysics.

CONTENTS

Investigation of Generation of the Second Optical
 Harmonic in Molecular Crystals

V. D. Shigorin

Calculation by the WKB Method of Resonator Modes
for Lasers with an Active Medium

B. P. Kirsanov and A. M. Leontovich

INDUCED RADIATION OF RUBY AT
LOW TEMPERATURES

A. M. Leontovich and A. M. Mozharovskii

The free-running, passive Q-switching, and self-mode-locking regimes are studied experimentally for a ruby laser at temperatures near 100°K. Features of lasing associated with the possibility of obtaining large gains (up to 0.5 cm^{-1}) in this active medium are considered in detail.

CHAPTER 1

FREE-RUNNING REGIME

§ 1. Main Features of the Lasing Kinetics

Work investigating ruby lasers at low temperatures constitutes an insignificant fraction of the enormous amount of experimental work devoted to studying the free-running regime of ruby lasers. The papers [1-8] are devoted to a study of the lasing kinetics in the temperature range 77-110°K. In the papers [9-13] the spectral composition of emission was also studied. The special features of lasing associated with reduced width of the luminescence line from 12 to 0.3 cm^{-1} as the temperature drops from 300 to 77°K were already established in the first stages [14, 15].

The most characteristic feature of the lasing kinetics as compared with room temperature kinetics is the regularization of the lasing regimes accompanied by an increase of the frequency of the relaxation pulsations. Regular damped pulsations are characteristic of ruby lasers at low temperature, and occur in practically any resonator possessing a sufficiently high Q, regardless of its type [4-6]. It is interesting that with increasing losses in the resonator at a given level of pumping, the rate of damping of pulsations increases and a transition occurs to a smooth pulse regime [1, 3, 4, 12]. In the papers [1, 5] a transition from smooth pulsations to regular undamped pulsations was observed when the rate of pumping at the end of the lasing pulse was increased in a resonator with low Q. The envelope of the train of spikes in this case repeated the shape of the pump pulse. The lasing kinetics and the mode composition of lasing were not studied simultaneously in these papers.

An investigation of the lasing kinetics with the aid of a high-speed photoregister [7] revealed that the field distribution in the near and the far zones in general has a smoothed-out form and pronounced maxima, corresponding to excitation of individual transverse modes, are absent (with the exception of the first spikes after the start of lasing, in which individual

1

maxima are distinctly seen). This indicated that a large number of transverse modes were excited simultaneously.

In [2] the dependence of the energy and duration of lasing on the rate of pumping in a resonator with low Q was investigated. A reduction in the duration of lasing and saturation of the output energy when the pump energy was increased was observed. However, the authors did not succeed in finding a satisfactory explanation for this. It is also established in [2] that the threshold pump energy depends very slightly on the Q of the resonator. More detailed measurements of the lasing threshold in resonators with different Q were made in [8], where a very weak dependence of the threshold pump energy on the coefficients of reflection of the resonator mirrors was also observed.

The lasing spectrum of ruby lasers at low temperature as a rule has a width of the order of 0.1-0.2 cm^{-1} [9-12], which usually corresponds to lasing of two to four axial modes. Two components separated by a gap of 0.38 cm^{-1} can be clearly resolved in the lasing spectrum.

These components correspond to the transitions $^2E \to \frac{1}{2} {}^4A_2$ and $^2E \to \frac{3}{2} {}^4A_2$. A detailed theoretical and experimental investigation of the difference of the threshold times of these components was given in [10-12]. All the experimental papers observed a smooth decrease of the frequency during the lasing process due to heating of the active element by the pumping light. The size of the frequency shift varies from 0.2-0.4 cm^{-1} in different papers.

The main properties of lasing of ruby at low temperature, such as pulsations of the intensity of emission, energy characteristics of lasing, and their temperature dependence, are investigated theoretically on the basis of balance equations of the Statz−de Mars type [16]. Here it is necessary to take into account that narrowing of the luminescence line is accompanied by an increase in the induced emission section σ, since $\sigma \sim 1/\Delta_l$ (Δ_l is the width of the luminescence line). For identical values of the inversion density n, this has the result that the intensity of superradiance (enhanced spontaneous emission) of ruby at low temperature is substantially greater than at room temperature. Superradiance is an additional cause of burning out (saturation) of the population inversion and must be taken into account in the balance equations.

Let us consider a ruby rod of length l with cross section s and concentration $n_2 = \frac{1}{2}(n_0 + n)$ of atoms on the upper level (n_0 is the concentration of active centers). In 1 cm^3 of such a sample, $\frac{n_0 + n}{2T_1} \frac{O}{4\pi} g(\nu) d\nu$ photons are emitted into the solid angle O per second in the frequency interval $d\nu$, where the factor $g(\nu) = \frac{\Delta_l}{2\pi} \frac{1}{(\nu - \nu_l)^2 + (\Delta_l/2)^2}$ describes the shape of the line, ν is the center of the luminescence line, and T_1 is the spontaneous lifetime of the upper laser level. The number of photons emitted per second from both ends of the rod can be expressed [17] as

$\frac{1}{T_1} \frac{2s}{\pi^2 l^2} \frac{n_0 + n}{2l\sigma n} \int_{-\infty}^{\infty} (e^{\frac{l\sigma n}{1+\nu^2}} - 1) d\nu$. Superradiance decreases the numbers of particles on the upper

level at the same rate. Taking this term into account, the balance equations can be written as

$$\frac{d\Phi}{dt} = (c\sigma n - \gamma) \Phi, \tag{1.1a}$$

$$\frac{dn}{dt} = n_0 (\rho - \rho_0) - n (\rho + \rho_0 + 2\sigma\Phi). \tag{1.1b}$$

Here Φ is the density of the photon flux inside the resonator, $\gamma = 1/T = c\sigma n_{\text{thr}} = cK_{\text{thr}}$ the reciprocal damping time of radiation in the resonator, ρ the probability per second of excitation of an atom by the pump, $\rho_0 = \frac{1}{T_1} \left(1 + \frac{2s}{\pi^2 l^2} \frac{1}{2\sigma n l} \int_{-\infty}^{\infty} (e^{\frac{2l\sigma n}{1+\nu^2}} - 1) d\nu \right)$. The last quantity can be approximated by the expression $\rho_0 \simeq \frac{1}{T_1}\left(1 + \frac{2s}{\pi^2 l^2}\frac{\sqrt{\pi}e^{2n_{thr}\sigma l}}{\sqrt{(2n_{thr}\sigma l)^3 + 1.6e^{-2n_{thr}\sigma l}}}\right)$. Using this expression, the

threshold rate of pumping can be written as

$$\rho_{thr} = \rho_0\,\frac{1 + n_{thr}/n_0}{1 - n_{thr}/n_0} = \frac{1}{T_1}\left(1 + \frac{2s}{\pi^2 l^2}\frac{\sqrt{\pi}e^{2n_{thr}\sigma l}}{\sqrt{(2n_{thr}\sigma l)^3 + 1.6e^{-2n_{thr}\sigma l}}}\right)\frac{1 + n_{thr}/n_0}{1 - n_{thr}/n_0} = \frac{1}{T_1}(1 + \delta(n_{thr}))\frac{1 + n_{thr}/n_0}{1 - n_{thr}/n_0}\,. \quad (1.2)$$

This rate depends exponentially on the threshold value n_{thr} of the inversion and increases rapidly with the latter, which may affect the lasing kinetics.

Like the ordinary balance equations, system (1.1) cannot be solved analytically. If for small deviations from the stationary level we neglect the nonlinear terms, only solutions of oscillatory type are obtained, having the form of damped pulsations whose frequency N and damping coefficient $1/t_{damp}$ can be written approximately as

$$N = \frac{1}{2\pi}\sqrt{c\sigma(n_0 - n_{thr}(\rho - \rho_{thr})}, \qquad (1.3)$$

and

$$1/t_{damp} = \frac{1}{2}\left\{\frac{n_0}{n_{thr}}(\rho - \rho_0) + \frac{\delta(n_{thr})}{T_1}\left(\frac{n_0}{n_{thr}} + 1\right)(2n_{thr}\,l\sigma - 1.5)\right\}. \qquad (1.4)$$

It is seen from (1.2)-(1.4) how the superradiance affects the lasing characteristics. The threshold pumping at large losses increases very rapidly (exponentially) with the losses, which will also affect the frequency of the spikes. The damping of the pulsations will also change. If the damping at first decreases with increasing losses, then starting with some level (depending on the rate of pumping) it increases (Fig. 1).

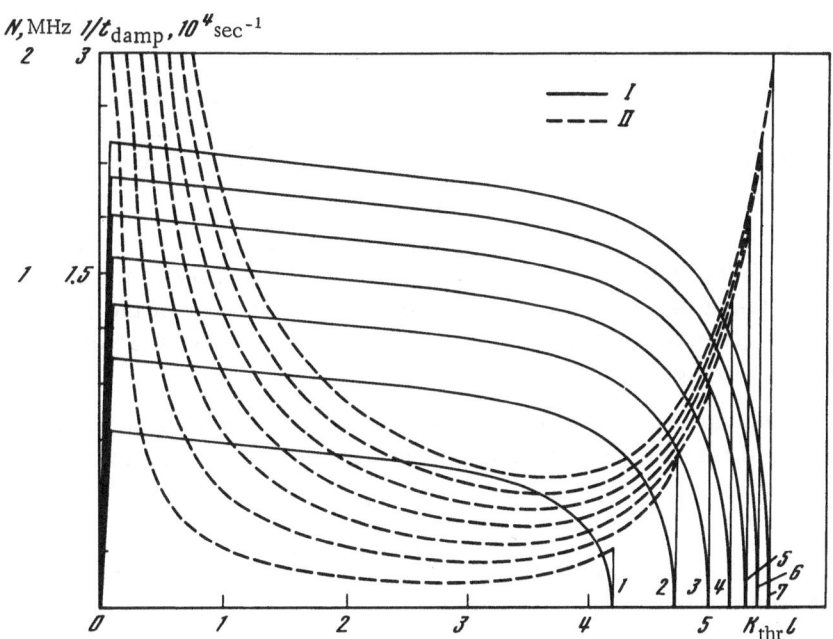

Fig. 1. Dependence of the frequency N (I) and damping co-efficient $1/t_{damp}$ (II) of pulsations on the losses in the reso-nator. Width of gain line for ruby, 2 cm^{-1}. Rate of pumping for curves 1-7 changes from $2T_1^{-1}$ to $8T_1^{-1}$.

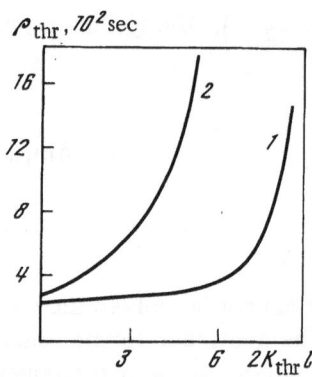

$\rho_{\text{thr}}, 10^2 \text{sec}$

Fig. 2. Dependence of the threshold pumping ρ_{thr} on $2K_{\text{thr}}l$ for a ruby of length 12 cm and cross section 1 cm². 1) 100°K, 2) 300°K.

However, it should be borne in mind that this effect plays a role only at large losses when the argument $2\sigma n_{\text{thr}}l = 2K_{\text{thr}}l$ in the exponential (K_{thr} is the threshold gain) is large. Figures 1 and 2 give as an example graphs of the dependences of ρ_{thr}, $1/t_{\text{damp}}$, N on the quantity $2K_{\text{thr}}l$ for a ruby of length $l = 12$ cm and cross section s = 1 cm². The specific form of the graph depends essentially on the value of σ, which for given losses determines the value n_{thr}. The effective value of σ in turn depends on the width of the luminescence line. In the calculation for Δ_l the value 2.0 cm⁻¹ was taken, which is larger than the value measured in [14, 15]. This value corresponds to the width of the luminescence line of standard ruby crystals of medium quality, in which there is a small inhomogeneous broadening of the line shapes due to mechanical stresses. It is seen that superradiance manifests itself in ruby at low temperature only when $2K_{\text{thr}}l \simeq 8$, i.e., $K_{\text{thr}} \simeq 0.3$ cm⁻¹. Although such large losses are rarely encountered, it will be shown below that the phenomena considered are responsible for the effect of self-breakdown of lasing described in Sec. 4 of this chapter. It is also seen from the graph in Fig. 2 that in a wide range of values of K_{thr}, the threshold pumping in a ruby laser at low temperature depends very weakly on K_{thr}. This phenomenon, investigated experimentally in [8], has so far remained without a satisfactory explanation.

Superradiance starts to affect the damping coefficient in ruby with $l = 12$ cm and s = 1 cm² when $K_{\text{thr}}l > 2$ and when $\rho/\rho_0 > 4$ (cf. Fig. 1). Apparently, it is just this effect of superradiance which can explain the occurrence of the "smooth pulse" regime observed in ruby at low temperature [1, 3, 4, 12].

It must be remarked that an exact calculation of the superradiance requires the values of numerous factors such as, e.g., the shape of the sample, the quality of the lateral surfaces, etc., which have not been taken into account in our calculations. In view of the roughness of the model, the estimates given below are primarily of an illustrative character, and the critical value $2K_{\text{thr}}l \simeq 8$ given is clearly too high.

The simplest balance equations also make it possible to explain the effect observed in [18, 19], in which the threshold pumping energy decreases when the temperature of ruby is lowered. The theoretical dependence of the threshold energy on the temperature for small resonator losses can be obtained from the following arguments.

In accordance with the total energy balance equation, taking into account the Boltzmann redistribution between the levels $\bar{E}$ and 2A in the Cr^{3+} ion, we have

$$\frac{dn'}{dt} = \frac{n_0}{2}\left(\frac{2\rho}{1+e^{-\Delta/kT}} - \frac{1}{T_1}\right) - n'\left[\rho + \frac{1}{T_1} + 2\sigma\Phi\frac{3+e^{-\Delta/kT}}{2(1+e^{-\Delta/kT})}\right], \qquad (1.1b')$$

where $n' = \frac{n_2}{1+e^{-\Delta/kT}} - \frac{n_1}{2}$ is the population inversion between the level $\bar{E}$ and the ground level, and $\Delta = 29$ cm⁻¹ is the distance between the levels $\bar{E}$ and 2A. The population inversion has a more precise meaning in (1.1b') than in (1.1) and (1.2), where the inversion is understood to be the total inversion between the levels ^2E and $^4A_2$. Integrating (1.1b') approximately with respect to t before the time of start of lasing t_{thr} (when $\Phi = 0$), we obtain for the threshold integral $R = \int_0^{t_{thr}} \rho dt$ of the pump the approximate equation

$$R = \ln\frac{3 + e^{-\Delta/kT}}{2} - \ln\left[1 - \frac{1 + e^{-\Delta/kT}}{2}\left(\frac{K_{thr}}{K_0} + \frac{t_{thr}}{T_1 R}\left(1 + \frac{K_{thr}}{K_0}\right)\right)\right] - \frac{t_{thr}}{T_1}, \tag{1.5}$$

where $K_0 = \sigma n_0$. The dependence of R on temperature at constant pumping can be found from this equation. The integral R is clearly proportional to the pumping energy required for onset of lasing. Figure 3 shows the calculated dependences of the pump integral on temperature. Data on the temperature dependence of T_1 cited in [15] were used in the calculation. It is seen from these graphs that the reason for the reduction of the threshold energy as the temperature decreases is the increased population of the working level $\bar{E}$ as a result of Boltzmann redistribution and also a reduction of the spontaneous emission losses caused by an increase of the lifetime T_1. As for the effect of resonator losses on the threshold pump energy, this effect weakens rapidly as the temperature is reduced, since the quantity K_0 increases rapidly.

The following question concerns regularization of the lasing regime in ruby at low temperature. The problem of the kinetics of emission during free-running conditions in solid lasers, as is well known, has been extensively discussed in the literature [20-27]. In the papers [20, 26] it was shown that regularity of pulsations in lasers occurs when a large number of modes are excited simultaneously. Under these conditions, the field of emission is distributed uniformly over the whole sample, "burning out" of inversion also occurs uniformly, and fluctuations in the gain, which lead to irregularity of the kinetics, do not occur. It was shown in [7] that simultaneous excitation of a large number of modes does indeed occur in ruby at low temperature. In this context it should be ascertained how the change of the parameters of ruby which occurs as the temperature is lowered can facilitate an increase in the number of simultaneously generating modes. To this end the results of our paper [20] can be used, in which the balance equations of [16], with allowance for nonuniformity of the transverse distribution of population inversion, were used to obtain the condition

$$\zeta = \frac{\Delta\psi''}{1.16L}\sqrt{\frac{c}{(K_0 - K_{thr})\rho}} < 1 \tag{1.6}$$

for simultaneous lasing of several transverse modes. Here $\Delta\psi''$ are the discrimination losses, defined as the difference of the losses of two neighboring transverse modes, and L is the length of the resonator. The quantity ζ is characterized in [20] as a coherence parameter (for $\zeta > 1$ the laser emission is spatially coherent). The quantity $\Delta\psi''$ is determined primarily by the type of laser resonator, and also by the character of the transverse distribution of the population inversion. Condition (1.6) means that if the pump power ρ is sufficiently large, excitation of several transverse modes occurs in each spike and the lasing is regular. Since K_0

Fig. 3. Temperature dependence of the pump integral R normalized relative to the value of R at 295°K for various values of t_{thr}. Inhomogeneous width of gain line, 2 cm^{-1}. Losses in the resonator, 30% for a double pass.

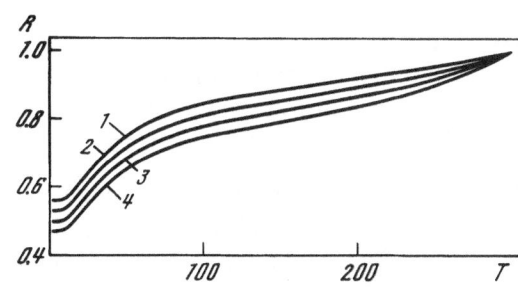

increases when the temperature is lowered, all other conditions being equal (fixed pump energy, definite type of resonator), the value of ζ in ruby at low temperature should be smaller than at room temperature, which facilitates regularization of pulsations.

In addition to regularization of the pulsations, it is also necessary to explain why they are not damped. The solution of this problem [21-27] involved certain well-known difficulties. Solutions having the form of undamped pulsations are not obtained in the framework of the balance equations in the single-mode model when no allowance is made for the spatial distribution of the field and inversion or its variation during the lasing process. In the general formulation of the problem [21], it is necessary to solve an infinite system of equations describing the mode fields. In addition, it is necessary to take into account that under actual conditions the laser parameters do not remain constant during the lasing process owing to heating of the active element by the pumping light, mechanical instabilities, burning out of inversion, etc. Because of all these difficulties, it seems that even in the general formulation of the problem, the theoretical investigations of the "spike problem" were limited as a rule to a consideration of two-mode generation. Thus, in [21], for example, the kinetics of laser emission were studied with allowance for two modes of an "empty" resonator and it is shown that undamped pulsations of radiation occur if the detuning frequency between the modes is close to a period of the pulsations. An analogous result was obtained in [22], where a computer calculation was carried out for multi-mode lasing kinetics for a laser with a δ-function transverse distribution of inversion. The regime of undamped pulsations is sometimes associated with the existence in the laser resonator of a medium with saturable absorption [23] or a nonlinear medium [24]. In [25], the undamped pulsations were explained by instabilities of the mode losses, and it was shown that such instabilities can occur as a result of mechanical vibrations of the laser components in conjunction with random discrimination of distinct modes in a laser with exterior mirrors.

All the above conjectures are physically plausible and in addition find experimental support. We can conclude from this that it is not possible to obtain an unambiguous answer to the problem in its most general form of the mechanism by which the undamped pulsations arise. Each of the mechanisms pointed out above may play a greater or lesser role, depending on the experimental conditions. However, they can all be completely or partly eliminated, and therefore in our view these conjectures explain only certain special aspects of the phenomenon. None of them permits an estimation of the possibility of occurrence of undamped pulsation regimes in specific lasers. In this connection an attempt was made in our paper [26] to determine more general conditions for obtaining regimes of various types, with allowance for change of the parameters of the active medium and laser resonator during the lasing process. The argument in [26] was based on a system of balance equations for two modes of a real resonator (filled with an active medium), each of which consisted of a superposition of modes of an "empty" resonator. Numerical solution of these equations on a computer showed that changes in the mode composition during the lasing process which arise due to slow modulation of the effective section σ by the mode frequency play an important role in determining the character of the lasing regime.

Slow modulation of σ can occur for the following reason. Heating of the active element by the pumping light causes the contour of the gain line to shift in frequency, and the rate of this shift does not usually coincide with the rate of change of the mode frequencies. For different modes, σ changes in different ways, and hence at a time when certain modes disappear from generation, others pass through the threshold conditions and start to generate.

In the computer calculation, the change of σ with time was approximated by a sinusoidal function with a "phase" shift of both modes. The "frequency" Ω in this function, which determines the rate of change of the sections σ, was chosen to correspond to the conditions in the

ruby laser. It depends on the difference between the rates at which the center of the luminescence line ν_l is shifted and the mode frequency $\nu_\mu = cq/2L$ (q is the axial mode index). Both
these rates are proportional to the rate of change of temperature of the active element and are
determined by the rate of pumping ρ, whence

$$\Omega = \frac{\sqrt{2}}{\ln 2} \frac{1}{\Delta_l}(D_l - D_\mu)\frac{n_0 h\nu Q}{C_v}\rho, \tag{1.7}$$

where C_v is the heat capacity per unit volume of the active medium, Q is the "heat yield" of the
pumping, equal to the ratio of the energy expended in heating the active element to the energy
used to excite atoms onto the upper laser level; $D_\mu = \frac{d\nu_\mu}{dT} = -\nu_\mu \frac{\varkappa + \alpha\,(\mu - 1)}{L/l + \mu - 1}$ (l the length of the
active rod, μ the index of refraction, α the thermal expansion coefficient, $\varkappa = d\mu/dT$).

Figure 4 shows some examples of the dependences calculated for laser emission as a
function of time.

On the basis of the calculation we may conclude the following. If both modes are initially
excited and then one of them drops below threshold conditions and ceases generation, the character of the lasing kinetics does not change. The intensity of the disappearing mode decreases mono-
tonically to zero. But if only one mode is initially generating and the other mode commences
generation after a certain time, then when the threshold is crossed, excitation of pulsations of
both modes occurs even if the first mode is quasistationary. Thus, the "appearance" of a new
mode in the lasing does not permit the pulsations to decay.

In real lasers a large number of modes are excited, which apparently does not change the
overall picture in principle, although it does make the situation more complicated. As lasing
takes place, the composition of modes changes constantly and when certain modes disappear
from generation, new modes continuously approach the lasing threshold. As we have seen, this
prevents the pulsations from decaying, so that undamped pulsations occur to the extent that heating of the active element continues and new modes enter generation. We note that the effect of
the appearance of new modes was investigated in detail in [27].

It is natural to assume that the character of the pulsations depends on how frequently new
modes enter generation and, in particular, how many modes enter generation during the lifetime
δt of a single pulsation. The larger this number, the greater is the probability of a regular regime, since when a large number of modes with arbitrary combinations of longitudinal and transverse indices are excited during the lifetime of a spike, burnout of inversion proceeds uniformly over the entire lasing volume. In addition, an increase of the number of modes entering generation should aid the development of the regime of undamped pulsations.

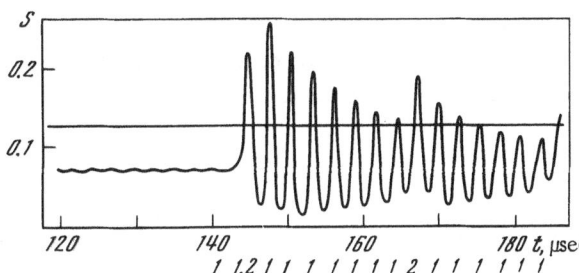

Fig. 4. Kinetics of two-mode lasing for a slow
variation of the sections $S_{1,2} = \sigma_{1,2}/\sigma = \frac{1}{2}$ (1 +
cos $\Omega t + \varphi_{1,2}$). σ, the section in the center of
the line: $\Omega = 3.32 \cdot 10^2$ sec^{-2}; $\varphi_1 - \varphi_2 = 0.7$.
Numbers under the spikes give the index of
the mode whose emission dominates in the
total intensity.

This conclusion is supported, e.g., by the results of experiments investigating lasers with a "running" active medium [28]. In that paper, regularization of the lasing regime and undamped oscillations were achieved by moving the active element with respect to the resonator mirrors. An investigation of the spectral composition of the emission from such a laser showed that a large number of axial modes are replaced in each pulsation.

Let us estimate the number of modes f which enter generation during the lifetime δt of a single pulsation. This number depends on the displacement of the gain line relative to the mode frequencies during the interval δt. Clearly, this displacement is tantamount to an increase of the effective gain for modes located on one of the wings of the gain line by an amount $\delta K = \left(\frac{dK}{d\nu}\right)_{\nu_\mu} \delta\nu$. The average value of $dK/d\nu$ for the modes participating in generation for a Lorentz gain line can be expressed approximately as $dK/d\nu = (\Delta_{las}/\Delta_l^2)K(\nu_l)$, where Δ_{las} is the width of the lasing spectrum. Using the fact that the transverse modes are degenerate with respect to losses, the multiplicity of degeneracy being approximately equal to the sum of the transverse mode indices, we can write f as $(\Delta_{las}^2/\Delta_l^2)K(\nu_l)(L^3/c\Delta\psi''^2)\delta K$. Substituting δK and expressing $\delta\nu$ in terms of experimental parameters as

$$\delta\nu = (D_l - D_\mu)\frac{n_0 h\nu Q}{C_\nu}\rho\delta t, \tag{1.8}$$

the final expression

$$f = \left(\frac{\Delta_{las}}{\Delta_l}\right)^4 \frac{L^3}{\Delta\psi''^2 c} K_{thr}^2 (D_l - D_\mu)\frac{2\pi h\nu Q}{C_v}\sqrt{\frac{n_0\rho}{c\sigma}} \tag{1.9}$$

can be obtained. In Eq. (1.8), $K(\nu_l)$ is replaced by K_{thr}, the threshold gain. The equation $\delta t = 2\pi/(n_0\sigma c\rho)^{1/2}$ [20, 26] has been used for δt.

Starting from a given width Δ_{las} of the lasing spectrum, the number of modes F participating in generation can be estimated analogously as

$$F = \left(\frac{\Delta_{las}}{\Delta_l}\right)^4 \frac{L^2}{\Delta\psi''^2} K_{thr}^2 \frac{\Delta_{las}2L}{c}. \tag{1.10}$$

From (1.9) and (1.10) we obtain

$$r = \frac{f}{F} = (D_l - D_\mu)\frac{\pi h\nu Q}{C_v \Delta_{las}}\sqrt{\frac{n_0\rho}{c\sigma}}. \tag{1.11}$$

As is easily seen, Eq. (1.11) can be rewritten as $r = \frac{1}{2}\,\delta\nu/\Delta_{las}$; i.e., the ratio of the number of modes entering generation during the lifetime of one pulsation to the number of generating modes is given by the ratio of the change of frequency over the lifetime of one pulsation to the width of the lasing spectrum. The width of the lasing spectrum can be determined directly from experiment. The formula

$$\Delta_{las} = 0.287\Delta_l\sqrt[4]{c(K_0 - K_{thr})(\rho - \rho_{thr})}\sqrt{K_{thr}c} \tag{1.12}$$

from [20]* may be used to estimate Δ_{las} theoretically during the first instants of generation in a laser with a homogeneously broadened gain line. Substituting (1.12) into (1.10), we get

$$F = 3.6 \frac{L}{c} \frac{\Delta_{las}}{\zeta^2}. \tag{1.13}$$

We attempt to determine the effect of f, F, and r on the character of the lasing regime. For $F \gg 1$, the distribution of the lasing emission field inside the resonator is homogeneous, as remarked above. Therefore, modes entering generation will have a large spatial overlap with the modes that are already generating, and their effect on the lasing kinetics will not be very great. Moreover, as a rule, when $F \gg 1$ the relation $r < 1$ holds.

If on the other hand F is small and $r \simeq 1$, there exist some modes among those commencing generation which have a small overlap with the already generating modes. This clearly leads to the appearance of undamped pulsations. The relation $r \simeq 1$ means that the mode composition is in fact reestablished completely during the lifetime of a single pulsation. This process is quite reminiscent of the process of Q-switching with many changing modes. For $F > 1$, $f > 1$, the character of the pulsations should be regular, since the development of lasing in each spike occurs under identical conditions. Apparently, this is the regime of lasing realized in a ruby laser in [6]. If on the other hand both F and f are of the order of unity, we get irregular, undamped pulsations, since burnout of inversion in each spike occurs nonuniformly and hence the initial conditions vary from spike to spike.

For $f = 0$ one would expect a regime of damped pulsations to occur. However, in practice it is not possible to achieve this situation. The condition $f = 0$ comes closest to realization in lasers with discrimination of transverse and axial modes. The emission of lasers of this type in general consists of several trains of damped pulsations [29]. Moreover, as in the case depicted in Fig. 4, the beginning of each successive train coincides with an instant when the lasing frequency jumps discontinuously, which corresponds to a change in the axial mode index by unity. Such a regime can be characterized as unstable.

Summarizing the arguments given above, the following scheme may be proposed which establishes a correspondence between the character of lasing and the values of the parameters F, f, and r:

1) $F \gg 1$, $r < 1$, regular damped pulsations;
2) $F > 1$, $r \simeq 1$, regular undamped pulsations;
3) $F \gtrsim 1$, $r \simeq 1$, irregular undamped pulsations;
4) $F \gtrsim 1$, $r < 1$, irregular damped pulsations;
5) $F \simeq 1$, $r \simeq 0$, unstable regime.

---

*In [20] the equation for the spectral width Δ_{las} has a somewhat different form and is expressed in terms of n_{thr}, σ, n_0, and ρ as

$$\Delta_{las} = \Delta_l \frac{0.37}{\sqrt{1 + 0.25 K_{thr} \, c / \sqrt{c(K_0 - K_{thr})(\rho - \rho_{thr})}}}.$$

The difference between this equation and (1.12) is due to the fact that in [20] it was assumed that all modes whose threshold times do not exceed the threshold time of the mode with maximal coefficient by more than the "burnout" time of inversion in the spike enter generation. Here, however, it was assumed that the intensities of the modes differ by a factor of 2 in the line center and on the boundaries of the line at the instant of maximum population inversion.

It is quite difficult on the basis of the relations obtained to reach an unambiguous conclusion about whether lowering the temperature of ruby facilitates development of undamped pulsations. One can only note that, of the quantities appearing in Eq. (1.11), those subject to the greatest variation are D_1 (0.14 cm^{-1}/deg at 300°K and 0.05 cm^{-1}/deg at 100°K), C_v (0.18 cal/g·deg at 300°K and 0.03 cal/g·deg at 100°K), as well as Δ_{las} and σ (as noted above). A relative estimate of the factor r in a ruby laser at low and at room temperatures shows that r increases by a factor of 5 as the temperature of ruby drops from 300 to 100°K. More precise calculations require that we be given the parameters of a specific laser.

2. Experimental Setup and Method of Measurement

Figure 5 shows a diagram of the experimental setup for investigating ruby lasers at low temperature. In our variant, a ruby crystal with ends cut off at the Brewster angle held by brass holders at the ends was placed in a glass tube. The ends of the tube and crystal were inserted into ebonite holders (since the coefficient of thermal expansion of ebonite is significantly larger than for glass and brass, the components were prepared with an interstitial gap of 0.05–0.1 mm). The device was placed in a tight-fitting double-elliptical illuminator with two IFP 2000 or IFP 5000 lamps, depending on the length of the crystal, and was fastened to the bottom of the illuminator by means of screwable ebonite handles. In order to prevent burning of the ebonite components by the pumping light, they were wrapped in aluminum foil.

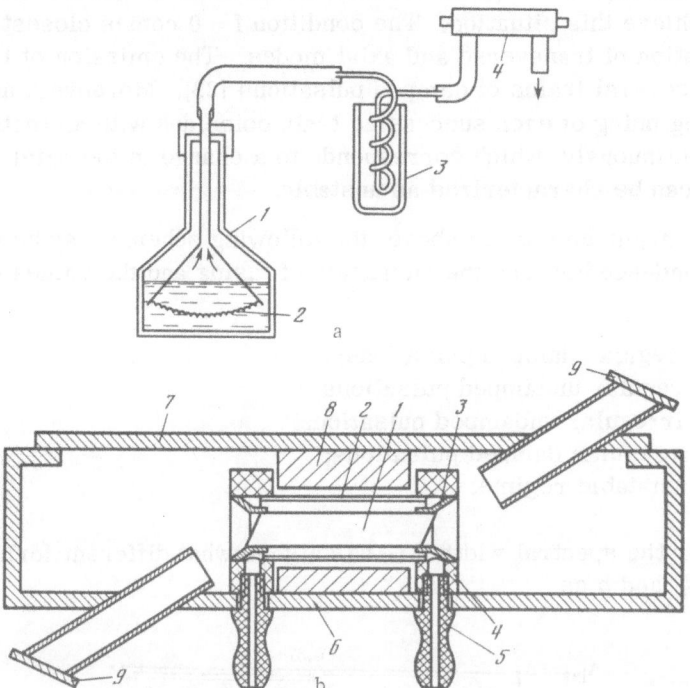

Fig. 5. Diagram of experimental setup. a) Nitrogen supply system: 1) ASD-15 Dewar; 2) heating coil; 3) Dewar, 2 liters; 4) illuminator: b) position of crystal in illuminator: 1) crystal; 2) glass tube; 3) brass holders; 4) ebonite holders; 5) ebonite handles; 6) housing of illuminator; 7) lid of illuminator; 8) reflector; 9) output windows.

The ruby was cooled by gaseous nitrogen evaporated from an ASD-15 Dewar in which a heating coil was placed. The supply of nitrogen could be regulated by changing the voltage on the heating coil. Nitrogen vapor then passed through a coil immersed in liquid nitrogen, after which it entered the illuminator. The supply hoses were covered with thermally insulating material. The temperature was monitored with the aid of a copper–constantan thermocouple. Eight to ten minutes after the system was turned on, the temperature of the ruby fell to −180°C, after which it remained constant. Heating of the ruby by the pumping light for a maximum pump energy of 10 kJ did not exceed 15-20°, and after 1 min the temperature returned to the initial value.

Since all the components of the tubing in contact with the illuminator were prepared from thermally insulating material, the temperature of the illuminator remained constant over a long period of time and there was no fogging of the output windows. The choice of materials with differing thermal expansion coefficients was also of great importance. Indeed, in the first instants of operation of the device, some of the nitrogen escaped through the gaps between the components into the inner cavity of the illuminator. Thus the illuminator was filled with dry nitrogen, which prevented fogging of the ends of the ruby and walls of the glass tube.

The lamps were powered from a capacitor bank of capacitance 1200 μF. The pumping pulse duration was approximately 1 msec. In all the experiments, the optical axis c of the ruby was located in a plane passing through the pumping lamps.

As is shown in [20, 26], the analysis of experimental results on the kinetics of free lasing requires a knowledge of the rate of pumping ρ and the distribution of population inversion in a cross section of the ruby crystal, since these parameters to a large extent determine the character of the lasing. The rate of pumping ρ (probability of excitation of an atom) in the case of a regular regime, which is what occurred in most of our experiments, can be expressed in terms of the frequency of the lasing pulsations,

$$\rho = \frac{1}{T_1} + \frac{4\pi^2 N^2}{c(K_0 - K_{\text{thr}})}. \tag{1.14}$$

Figure 6 shows experimental results giving ρ as a function of the pump energy E for the case of a ruby 120 mm long and 12 mm in diameter pumped by two IFP 2000 lamps.

The distribution of the population of excited atoms of chromium and the gain were found by photometric scanning of photographs of the distribution of luminescence over the end surface of the ruby. The photographs were obtained by frame-by-frame scanning using the SFR-2M device. Ruby was used which had plane-parallel ends and a transparent rough lateral surface without mirrors.

In calculating the inversion distribution, allowance was made for the fact that owing to the finiteness of the viewing angle of the optical system, the intensity of luminescence at each

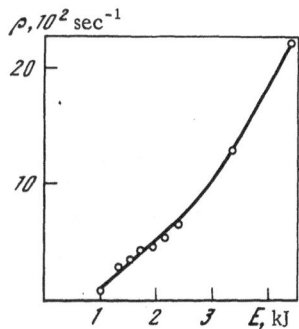

Fig. 6. Dependence of rate of pumping ρ on pump energy E.

point of the end surface of the crystal describes the inversion averaged over some region of the the cross section. The size of this region is of the order of θl (θ is the viewing angle of the optical system). The angle θ can be measured if we bear in mind that in the central part of the cross section of the rod, the distribution of the gain is in general well approximated by a parabola,

$$K(x) = K_1 + \chi x^2, \tag{1.15}$$

where K_1 is the gain on the axis of the rod and x is the distance from the axis. The intensity of luminescence in this case is described by the expression

$$I(x) \sim \int_0^l dz \int_0^{\theta/2} \gamma d\gamma \int_0^{2\pi} n_2(x) e^{\left(K_1 z - \chi x^2 z - \frac{\chi \gamma^2 z^3}{3} + \chi x \gamma z^2 \cos\varphi\right)} d\varphi. \tag{1.16}$$

In our experimental setup, $\theta \simeq 1°$ and a numerical calculation shows that the contribution from the terms containing θ does not exceed 3–4% of the total intensity.

When terms containing θ are neglected, the formula describing the ratio of the intensities of luminescence emitted by two points of the end surface located a distance x_1 and x_2 from the axis looks like

$$\frac{I(x_1)}{I(x_2)} = \frac{K_0 + K(x_1)}{K_0 + K(x_2)} \frac{K(x_2)}{K(x_1)} \frac{e^{K(x_1)l} - 1}{e^{K(x_2)l} - 1}, \tag{1.17}$$

where K_0 is the absorption coefficient of the unexcited medium. The distribution K(x) of the gain was found by numerical solution of Eq. (1.17). Figure 7 shows the distribution K(x) so obtained for pump energy equal to 3.1 kJ at different times. The value K(0) at time immediately preceding lasing was taken equal to 0.22 cm^{-1} from the threshold conditions (the reflection coefficient of the end of the ruby was 7.6%). The values of the pump power ρ determined from the graphs in Fig. 7 differ by not more than a factor of 2 from the values calculated in terms of the frequency of the laser pulsations. With the method of measurement used, this is completely satisfactory. The coefficient χ in the quadratic term in Eq. (1.15) for the maximum of the distribution (curve 4) is 1.2 cm^{-1}/cm^2.

3. Spatial Characteristics of Emission, Lasing Spectra, and Their Dynamics. Results and Discussion

The main purpose of the experiments described in this section was to ascertain the relation between the character of lasing in a plane-parallel resonator, the form of the transverse distribution of the gain, and the dynamics of the lasing spectrum. The measurements were made on ruby crystals of four types with chromium concentration 0.05%.

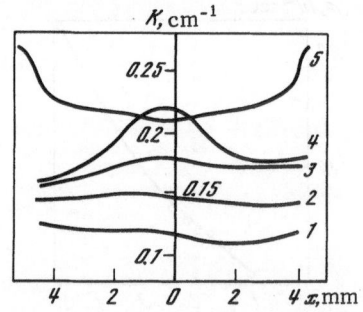

Fig. 7. Distribution of the gain along the longitudinal coordinate x. Pump energy 3.1 kJ. 1) After 128 μsec prior to start of lasing; 2-4) successive moments of time spaced 32 μsec apart; 5) 32 μsec after end of lasing.

Crystal A had l = 120 mm, diameter 12 mm, and a rough polished lateral surface. Dielectric mirrors with reflection coefficients R_1 = 30 and R_2 = 90% were placed at the ends.

Crystal B had l = 65 mm, diameter 13.5 mm, and a polished lateral surface with end mirrors with R_1 = 30 and R_2 = 100%.

Crystal C had l = 120 mm, diameter 12 mm, and a rough polished lateral surface without reflecting coverings.

Crystal D had l = 65 mm, diameter 13.5 mm, and a polished lateral surface with R_1 = R_2 = 95%.

Crystals A, B, and C had plane-parallel ends. The ends of crystal D were cut off at an angle 30' relative to each other (so that the angle between the mirrors was also 30').

The spatial distribution of emission and of the lasing spectra at 90°K were studied by an SFR-2M slit scanner.

It is clearly seen from the scans of the far zone of the emission of rubies of the various types, shown in Figs. 8-10, that the form of the scans changes as compared to room temperature. The pulsations of the emission become almost regular and are often barely discernible. The emission distribution sometimes becomes continuous, without structural spots; this situa-

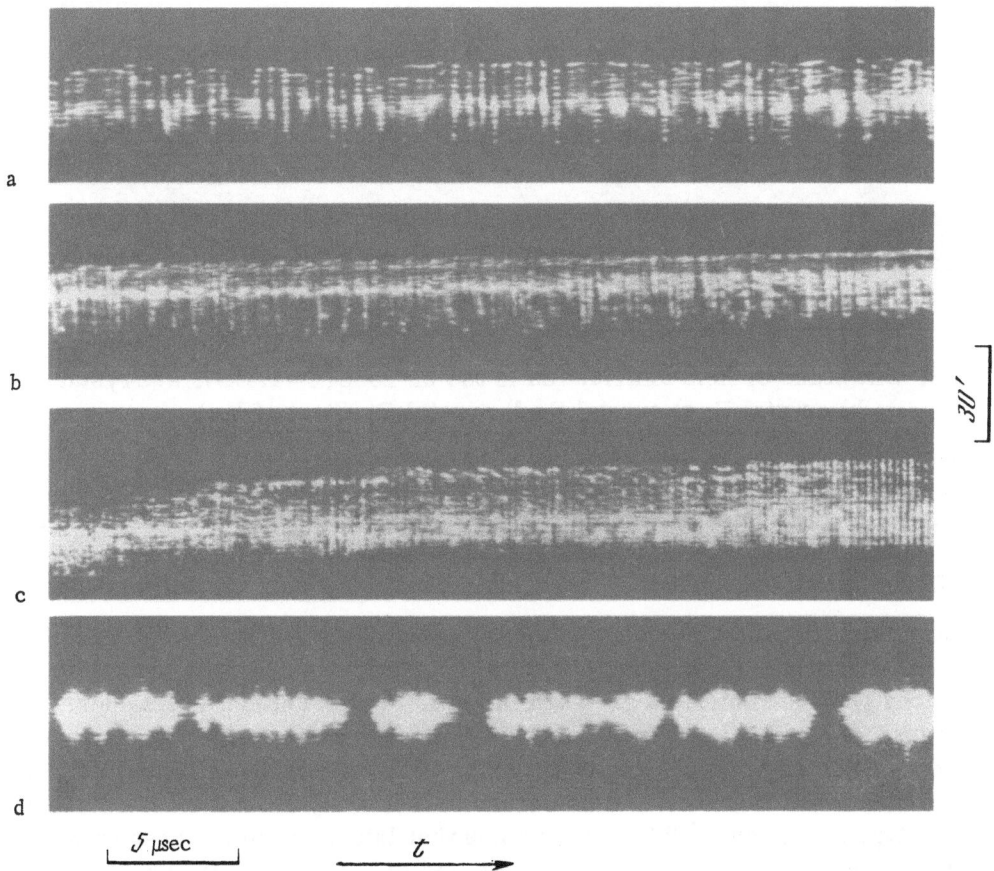

Fig. 8. Slit scans of the field distribution in the far zone for crystal C (30 sec after onset of lasing). a) E = 1.2 kJ; b) E = 2.4 kJ; c) E = 4.3 kJ; d) room temperature, E = 4 kJ.

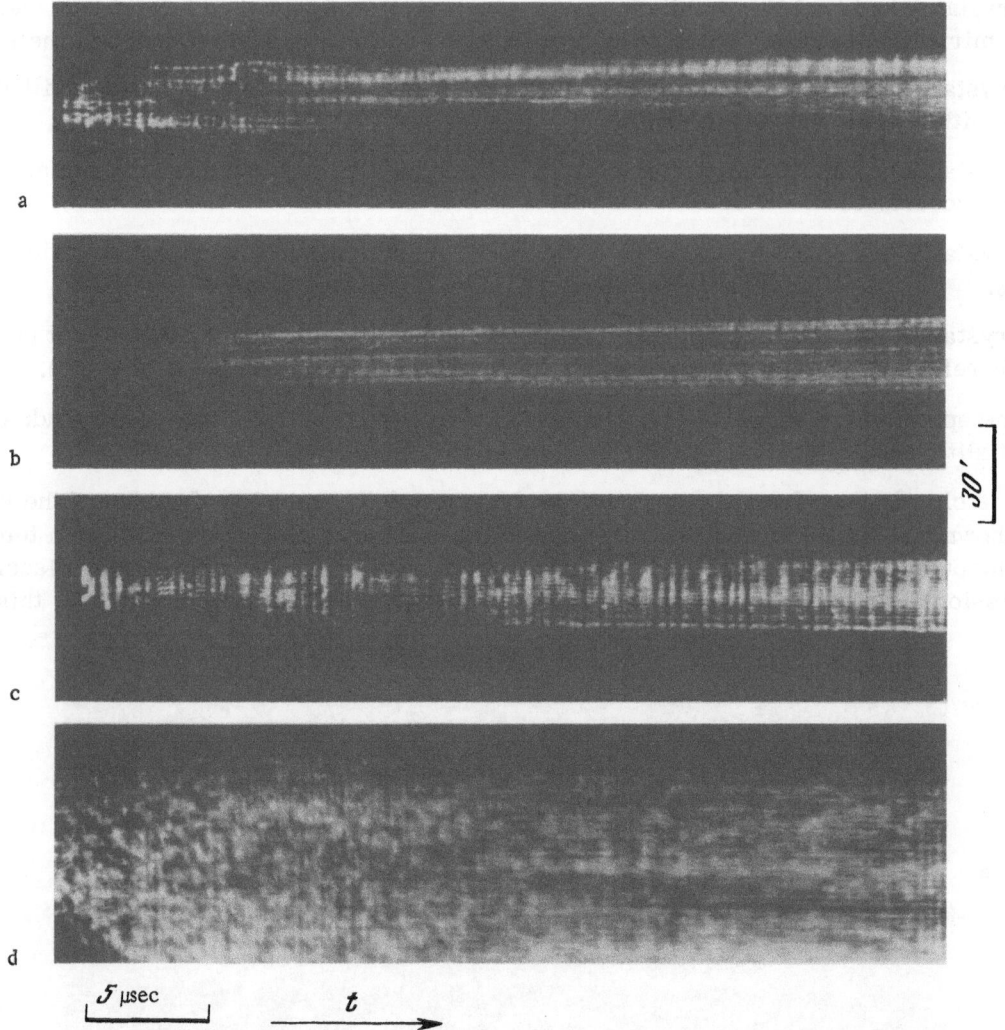

Fig. 9. Slit scans of field distribution in the far zone (E = 4.3 kJ). a) Crystal A,
b) crystal B; c) crystal C, d) crystal D. Start of lasing.

tion is usually seen on the quasistationary portions of the scan. The tendency to regularization of the regime when ρ increases can be followed most distinctly in Fig. 8, which shows slit scans for different pump energies.

The fact is remarkable that right at the start of lasing (cf. Fig. 9), when judging from the frequency of the pulsations the pumping power is a maximum, there is no complete regularization of pulsations and sharp maxima are seen in the field distribution, which indicates simultaneous excitation of a small number of transverse modes. Subsequently, the character of the dynamics changes; for crystals of type A, smearing out of the sharp maxima in the field distribution and regularization of the kinetics occurs immediately after the onset of lasing, while for crystals of types B, C, and D this occurs somewhat later, and in the case of crystals of type D, individual maxima can be seen even at the end of lasing. The frequency of the pulsations of lasing in crystals A and B is greater than in crystals C and D. The pulsations of the emission in crystal D are much more pronounced than in the other crystals and are not damped even at the end of the lasing process.

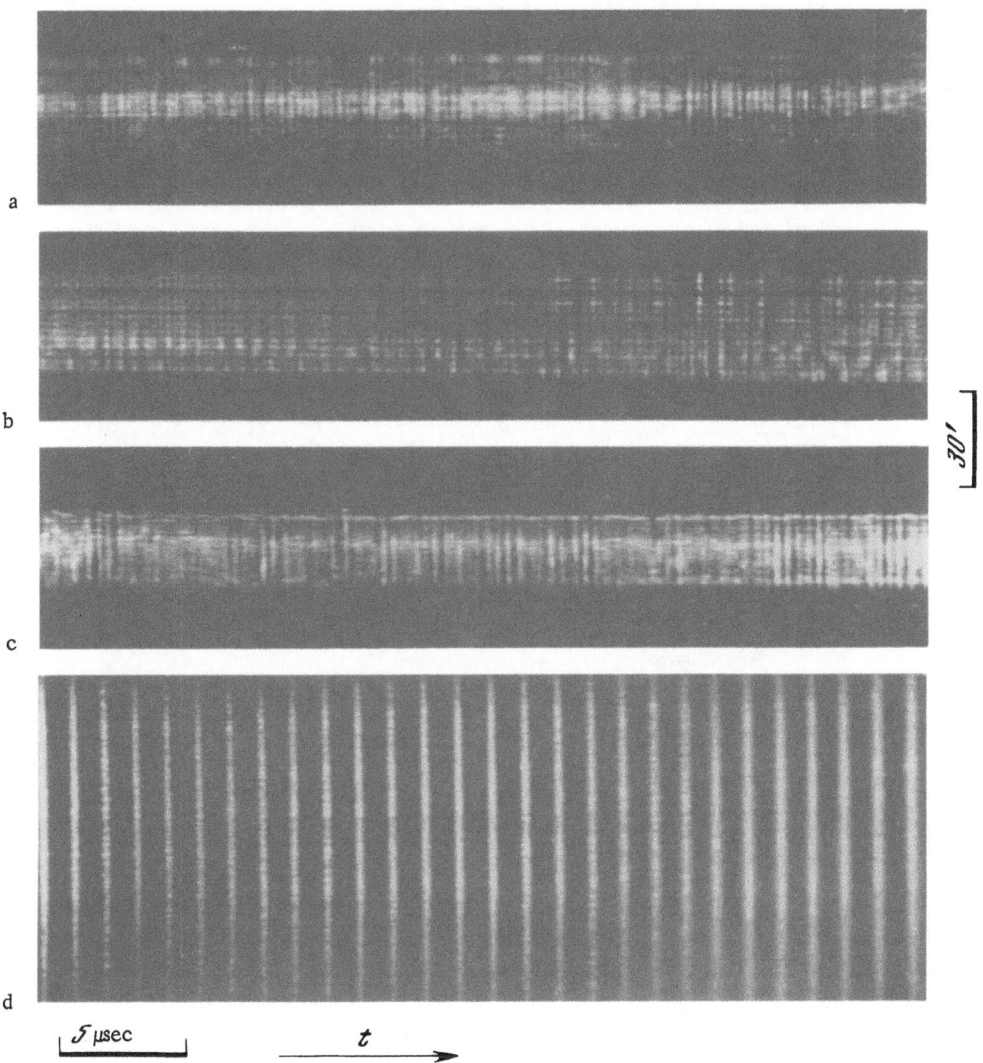

Fig. 10. Slit scans of field distribution in the far zone (E = 4.3 kJ). a) Crystal A; b) crystal B; c) crystal C; d) crystal D. a–c, taken 30 μsec after onset of lasing; d, 100 μsec after onset of lasing.

An interesting effect observed during the free-running mode of a ruby laser at low temperature was a drift in the direction of maximal intensity of emission (Figs. 8-10). The average value of the deviation of the light beam was approximately 5, 4, and 3' respectively for crystals A, B, and C. The reason for this phenomenon remains obscure. It can only be hypothesized that the drift in the direction of maximal intensity is associated with inhomogeneous heating of the crystal and formation of thermal inhomogeneities of the wedge type. As follows from the papers [30, 31] on asymmetric resonators, in this case the output beam should deviate in a direction opposite to the vertex of the wedge. This is confirmed by data on the angular distribution of emission from crystal D, for which the maximum in the intensity distribution deviates by an angle 15-20' from the crystal axis.

The emission spectrum was studied using a Fabry–Perot IT-51-30 interferometer with bases of 10 and 25 mm. The resolution capability of the interferometer was verified using an

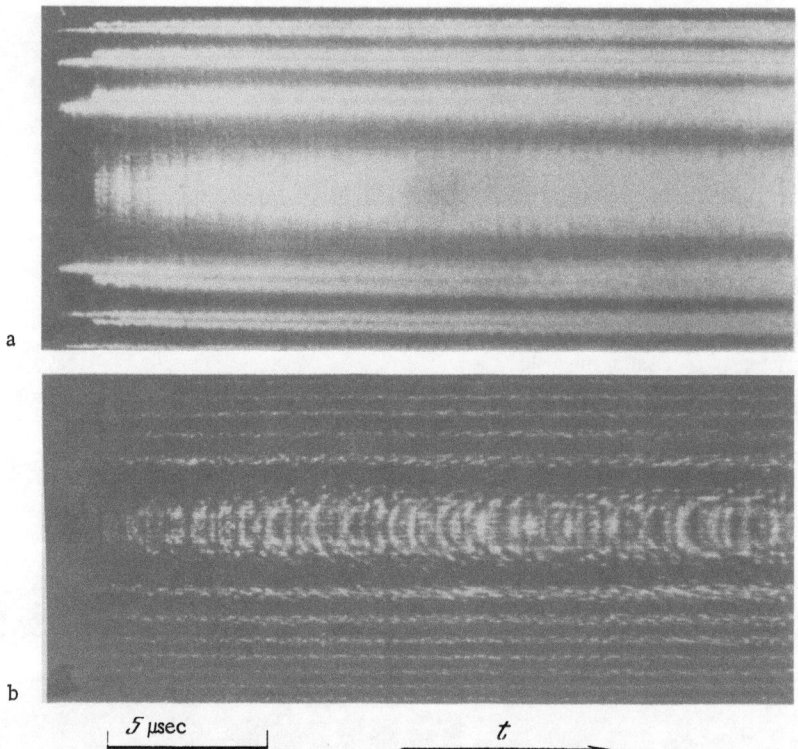

a

b

$\overline{\quad 5 \,\mu\text{sec} \quad}$ $t \longrightarrow$

Fig. 11. Slit scans of the lasing interference patterns for lasing
by crystal A (E = 4.3 kJ). a) Base of etalon 10 mm, start of
lasing; b) base of etalon 25 mm, start of lasing.

LG-75 gas laser and was $\frac{1}{12}$ of the region of dispersion. This was sufficient to resolve neighboring axial modes ($\Delta(\nu/c) = 0.041$ cm^{-1}).

Figure 11 shows slit scans of the interference patterns for crystal A. The splitting of the lower level $^4A_2$ is visible and equal to 0.38 cm^{-1}. For the transition $^2E \rightarrow \frac{3}{2}\,^4A_2$ occurring at shorter wavelengths, the threshold is reached earlier, which agrees qualitatively with the results of [10, 12]. The difference between the threshold times is approximately 1 μsec. The maximal width 0.25-0.30 cm^{-1} of the spectrum is attained 20-25 μsec after the onset of lasing. In each spike, 4-5 axial modes generate simultaneously, and the set of modes changes from spike to spike. Under quasistationary lasing conditions the spectrum has the form of an almost continuous band of width 0.1-0.2 cm^{-1}. This indicates that lasing takes place by many transverse modes. The average frequency of lasing decreases smoothly with time. As already mentioned above, this change is associated with a shift in the center of the luminescence line due to heating of the crystal by the pumping light. The center of the luminescence line is displaced at the temperature of the nitrogen at a rate $d(\nu/c)/dT = 0.05$ cm^{-1}/deg [19]. Thus, when the shift of the frequency during lasing is known, it is possible to determine the increase of the temperature of the crystal. For example, at a pump energy of 4.3 kJ, the crystal heats up by 12° during the lasing process.

Using Eq. (1.8), it is possible to estimate from the known rate of displacement of the center of the lasing line the "thermal yield" Q, which is equal numerically to the ratio of the energy of a flash of the pumping lamp used up in heating the crystal to the energy which goes to create population inversion. The average value of Q calculated with the aid of the lasing interference patterns was equal to 5.

We discuss the above experimental results taking into account Eqs. (1.9)-(1.11) which determine the values of the factors F, f, and r. We note right away that for crystals of type A and B, the value of r calculated from (1.11) or determined directly from the lasing interference patterns does not exceed 0.1. The mode composition of lasing is quasistable, and therefore in this case the lasing pulsations should decay, which is in fact observed experimentally. For crystal D, r = 0.4. Thus the mode composition changes to a significant degree within the confines of a single spike. The change of the mode composition can be seen quite well in the slit scan shown in Fig. 10d. It is seen from this figure that the set of sharp maxima in the angular distribution of the field, which corresponds to lasing by a definite combination of transverse modes, changes from spike to spike. The fact that a regime of undamped pulsations is realized for crystal D accords well with the model proposed in the preceding section, in which reestablishment of the mode composition should facilitate the development of undamped pulsations.

The difference in the values of r for crystals A, B, and C is due to the different losses in these crystals. The large losses in crystal D caused by the nonparallel mirrors results in narrowing of the lasing spectrum [in accordance with Eq. (1.12)] by a factor of 3-4 compared with crystals A and B. Since this factor is proportional to the ratio of the rate of shift of the luminescence line to the width of the lasing spectrum, the value of r for crystal D exceeds by a corresponding factor the value for crystals A and B.

As for crystal C, the value of r is approximately 0.5. However, in this case it is not possible to follow the development of the pulsations since the lasing only lasts for 50-100 μsec, after which it shuts off abruptly. This phenomenon is discussed in more detail in the following section.

Regularization of lasing during the lasing process can be related to an increase of the number F of modes participating in lasing. The increase in F during the lasing process can be judged with sufficient definiteness in terms of the broadening of the width of the lasing spectrum, which in turn corresponds to an increase in the number of axial modes participating in lasing, and in terms of equalization of the transverse distribution of the lasing field, which indicates an increase in the number of transverse modes. In order to ascertain the reasons for an increase in the number of transverse modes participating in lasing, it clearly makes sense to study the change of the transverse distribution of the gain during the lasing process. We recall that the character of this distribution has an important effect on the value of the discrimination losses (Eq. (10) of [20]), which according to (1.6) determine the coherence parameter ζ. At the same time, a theoretical estimate shows that F changes in inverse proportion to ζ^2 [Eq. (1.13)]. Thus, equalization of the transverse distribution K(x) during the lasing process, which occurs as a consequence of predominant burning out of inversion at the points of the cross section of the rod where the inversion is maximal, may lead to a significant increase of the number of generating transverse modes.

As is seen from Fig. 7, at the start of lasing K(x) is characterized by large values of the quadratic coefficient χ. An estimate of F under experimental conditions corresponding to Fig. 7 gives a value of the order of 2-3, in fair agreement with the value estimated from data on the angular structure and spectrum of emission. Quantitative estimates of the change in F during the lasing process are quite difficult, since the form of the distribution K(x) is not known. However, the qualitative agreement of the data on the increase of the number of generating modes with the proposed model is entirely satisfactory.

We note that the development of the central maximum of K(x) takes place after negative absorption is attained in the medium and the peripheral regions of the crystal have been made partly transluscent by the pumping light. Since the threshold gain in crystal A is much less than in crystals C and D, the initial distribution K(x) in crystal A should be smoother. In crystals B and D the distribution K(x) should be more peaked due to focusing of the pumping light by the polished lateral surfaces. During the lasing process, K(x) becomes smoothed out due to rapid burnout of the inversion in the central region of the rod. This leads to smaller discrimination losses for the transverse modes [20]. Starting at a certain time, lasing by a large number of transverse modes becomes possible, which leads to complete regularization of the lasing regime. Naturally, the transition to a regular regime occurs much faster in crystal A than in crystals of the other types.

The decrease of the frequency of pulsations for crystals C and D is due to the increased threshold pump power [Eq. (1.3)].

4. Phenomenon of Self-Breakdown of Free Lasing

The phenomenon of self-breakdown of free lasing in a resonator with low Q, i.e., shortening of its duration when the pump energy is increased and saturation of the emission energy, was previously observed in ruby lasers at low [2] and at room temperatures [32, 33].

In our experiments, the duration of free lasing and the energy of emission were investigated simultaneously by considering the distribution K(x) and its variation during the lasing process. Crystals of type C were used in the measurements.

The duration of lasing t_{las} was measured as a function of the pump energy E using slit scans taken on the SFR-2M. It was discovered that the duration of lasing decreases from 258 to 58 μsec when the pumping energy increases from 1.0 to 4.3 kJ. At large pumping energies, t_{las} changes inversely to E. The emission energy W was measured using IMO-1 and IZhK-1 devices. The energy first increases linearly with the pump energy and then starts to flatten out, reaching a stationary value of 0.67 J. The experimental results on the duration of lasing and the output energy are shown in Fig. 12.

It is important that at the time of breakdown of lasing, pumping still continues and drops altogether by a factor of 1.5 relative to the maximal value. This was established by successively recording two slit scans on the same piece of film. In the first slit scan, emission was scanned from a ruby in a resonator formed by the ends of a crystal without reflecting coverings. In the second, the emission of the same crystal was scanned in a semiconfocal resonator. It turned out that lasing ceases suddenly, almost without any decrease in the frequency of the pulsations. If we attempt to explain this phenomenon by an abrupt increase in the losses in the resonator, it is clear upon comparing the two scans that the existence of these additional losses depends essentially on the shape of the resonator. On the other hand, it is known [20] that whereas the

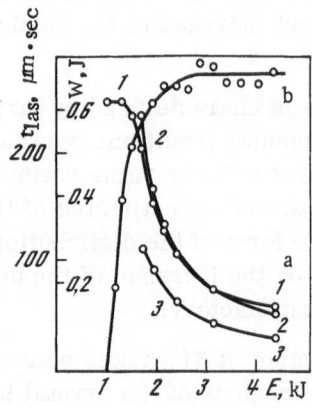

Fig. 12. Dependence of t_{las} (a) and W (b) on E. 1) Experimental; 2) calculated from Eq. (1.20) with ρ determined from the frequency of the lasing pulse; 3) calculated from Eq. (1.20) with ρ determined from graphs of K(x) at different times.

losses in stable resonators are determined solely by the output radiation in front of the mirror, in unstable resonators and resonators with a near-critical configuration (plane-parallel, concentric) the losses are strongly dependent on the transverse profile of the inversion. In this connection we considered the variation of the distribution K(x) during the lasing process.

It is seen from Fig. 7 that even 64 μsec prior to the onset of lasing, the inversion is not concentrated at the axis of the rod because of strong absorption of the pumping light by the peripheral region. Therefore, the resonator continues to be unstable almost up to the very start of lasing, since the ruby itself corresponds to a negative lens with focal distance 600-800 cm [7]. Concentration of inversion near the axis takes place only shortly before onset of lasing, when the peripheral regions of the crystal are partly transluscent to the pumping light. Lasing begins in the region of maximum inversion, and then, as the pumping light causes the region in which the mean population inversion attains the threshold value to get larger, the size of the lasing region increases, but does not exceed approximately 0.7 times the diameter of the crystal. Here the oscillations δK of the gain over the lifetime of a single pulsation are small compared with the initial inhomogeneities of K(x). Indeed, as shown in [20],

$$\delta K < \frac{2\pi \sigma n_0 (\rho - \rho_{thr})}{\sqrt{n_0 \sigma c (\rho - \rho_{thr})}} = 2\pi \sqrt{\frac{\sigma}{c} n_0 (\rho - \rho_{thr})} = \frac{4\pi^2 N}{c} \simeq 4 \cdot 10^{-3} \text{ cm}^{-1}.$$ At the same time, the initial

inhomogeneity of K(x) over a length $x \simeq 2$ mm is $\Delta K \simeq \chi x^2 \simeq 0.05$ cm^{-1}.

Thus, in the central lasing region, the inversion remains constant to within insignificant changes over the lifetime of a single spike. In the peripheral region the inversion continues to increase due to pumping. This leads to eventual reversal of the distribution of inversion, i.e., it remains approximately equal to the threshold value in the center, while at the boundaries it attains quite large values.

We show that for the distributions obtaining prior to the onset and after termination of lasing, the configuration of modes and their losses are essentially different, and the increase of the losses caused by transformation of the inversion profile during the lasing process is the fundamental effect responsible for the phenomenon described above.

According to [20], when the inversion has a maximum near the axis in an active medium, the lasing field can be confined in the region of this maximum. The condition for confinement of at least one mode imposes restrictions on the inhomogeneity of the permittivity $\Delta \varepsilon$ (x) = $\Delta \varepsilon'$ + $i\Delta \varepsilon''$ ($\Delta \varepsilon'$ (x) = $2\mu_0 \Delta\mu$ (x), $\Delta \varepsilon''$ (x) = $(\lambda/2\pi)\sigma \Delta n$ (x)), having for a dielectric of dimension a the form

$$a \frac{\nu}{2\pi c} \text{Re}(\sqrt{\Delta\varepsilon}) > \frac{\pi}{2}; \qquad \text{Re}(\sqrt{\Delta\varepsilon}) = [^1/_2 (\Delta\varepsilon' + \sqrt{\Delta\varepsilon'^2 + \Delta\varepsilon''^2})]^{1/2} \tag{1.18}$$

($\Delta\varepsilon$ is the value of the inhomogeneity over a distance a). In an unstable resonator $\Delta\varepsilon' < 0$ and $|\Delta\varepsilon''| \ll |\Delta\varepsilon'|$. Therefore, $\text{Re}(\sqrt{\Delta\varepsilon}) \simeq \frac{\Delta\varepsilon''}{\sqrt{-\Delta\varepsilon'}}$. The dimension a was found by slit-scanning the field in the near zone and was 0.6 cm in the first instants of lasing. The value of the inhomogeneity $\Delta\varepsilon''$ corresponding to a can be found from Fig. 7. It is equal to $2.1 \cdot 10^{-6}$. We can take for $\Delta\varepsilon'$ a value corresponding to the change of the index of refraction at radius 0.3 cm for a lens-shaped medium with focal length 600 cm. This value is $-1.5 \cdot 10^{-5}$ [7]. When these values are substituted into condition (1.18), inequality holds. This means that during the initial instants of lasing the field will be confined in the central region of the ruby. Therefore, mode losses due to diffraction at the boundaries of the reflecting surface play no role and the main factor determining the transverse mode losses is nonuniformity of the transverse distribution of inversion.

The situation changes in an essential way for the reversed distribution of inversion hold-ing after cessation of lasing. The lasing field will no longer be confined in the central region and therefore diffraction effects at the ends of the ruby start to play a role. In calculating the diffraction losses, the boundary of an unstable resonator (with a quadratic distribution $\Delta\varepsilon'$) can be approximated by a resonator with inclined mirrors (with a linear distribution $\Delta\varepsilon'$), and the formulas derived in [30] may be used. It is easy to show that the formula for the coeffi-cient of diffraction losses in the case of modes confined on one side by the boundaries of a diaphragm or mirrors and on the other, by a caustic, has the form

$$K_{diff} = 0.412\xi \sqrt{\nu/2\pi cl}, \tag{1.19}$$

where ξ is the angle of misalignment of the mirrors.

Since the gradient of the index of refraction for a ruby of medium quality is approximate-ly $3 \cdot 10^{-4}$ cm^{-1}, the corresponding value 4.10^{-3} was taken for ξ. In this case, calculation by Eq. (1.19) gives $K_{diff} = 0.14$ cm^{-1}. Thus, after the central maximum of the distribution of in-version is "smeared" out, the value of the threshold gain increases by not less than 0.14 cm^{-1} for a ruby of medium quality and becomes approximately equal to 0.36 cm^{-1}. As was shown in § 1, for such losses the effect of superradiance is already great, and the value of the threshold pumping increases so much that it begins to exceed the available pump power. Moreover, the dependence of ρ_{thr} on the losses is so steep that lasing terminates abruptly, almost without a smooth change in the frequency of pulsations.

It is obvious from the above arguments that the duration of lasing coincides with the time it takes to "smear out" the central maximum of the distribution of population inversion and is equal to

$$t_{las} = \frac{\Delta K}{K_0 (\rho - \rho_{thr})}. \tag{1.20}$$

ΔK is found experimentally from graphs of the distribution $K(x)$ (cf. Fig. 7). The pump power can also be found from a sequence of such graphs at different times.

The values of t_{las} calculated by Eq. (1.20) are shown in Fig. 12. The pump power in one case was calculated directly from graphs of $K(x)$ at different times, and in the other case in in terms of the frequency of the lasing pulsations [Eq. (1.14)]. As is seen from the figure, cal-culation using Eq. (1.20) gives good agreement with experiment when the threshold is substan-tially exceeded.

It is easy to obtain from (1.20) an expression for the output emission energy

$$W = \frac{s_{las}l}{2\sigma} h\nu\Delta K, \tag{1.21}$$

where s_{las} is the surface area of the region of lasing. Calculation using Eq. (1.21) gives a value 0.5 J for W, which for the method of measurement used is in good agreement with the maximal experimental value 0.67 J. In the calculation, σ was estimated using the formula

$$\sigma = \frac{\lambda^2}{4\pi^2\mu^2 T_1} \frac{1}{\Delta l} \tag{1.22}$$

for a Lorentz line. The value of Δ_{las} was found using an IT-51-30 Fabry–Perot interferom-eter and was equal to 2.0 cm^{-1}.

For near-threshold pumping, Eqs. (1.20) and (1.21) do not give agreement with experi-ment since in this case the duration of lasing is already determined by the time it takes for the

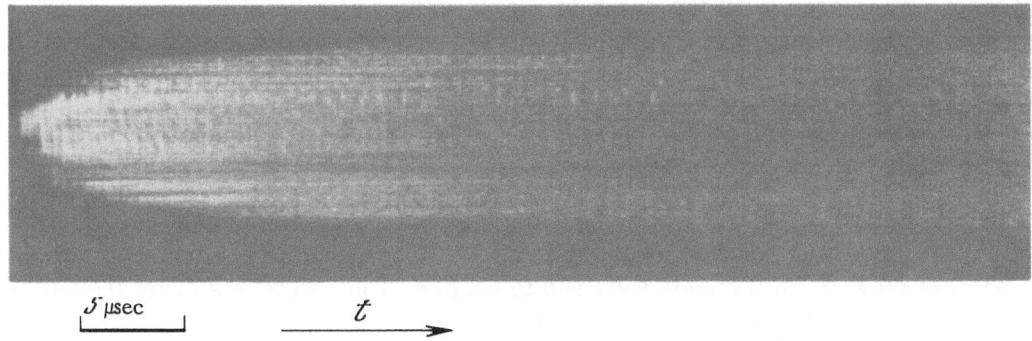

$\mathit{5}$ μsec t

Fig. 13. Slit scan of the distribution of the field in the near zone for crystal B (E = 4.3 kJ).

pump to decrease to the threshold level, which will be less than the time required to "smear out" the central maximum of the inversion distribution.

As already noted in the preceding section, smoothing out of the transverse distribution of inversion also affects lasing in a high-Q resonator. Figure 13 shows a slit scan of the emission from a crystal of type B. At the time of "smearing out" of the central maximum, lasing extends over the entire cross section of the crystal, its intensity dropping abruptly.

In order for this type of phenomena to occur also at room temperature, it is necessary that the value of $2K_0 l$ be sufficiently large (approximately equal to 10). This can occur for rubies of large length or with a high chromium concentration. We remark that ruby crystals with a high concentration of chromium are usually similar to a negative lens and do not become transformed during the lasing process into a positive lens, which as we have seen is necessary for self-breakdown of lasing to occur. It seems that the explanation for the effect of self-breakdown of lasing in ruby lasers at room temperature observed in [32, 33] should be sought in the effects of superradiance and the influence of the distribution of population inversion considered above, although owing to the irregular character of the lasing regime, these effects should be less pronounced.

CHAPTER 2

PASSIVE Q-SWITCHING REGIME

§1. Main Features of the Kinetics

The main properties of solid lasers with bleachable filters, which are the most common type of passive shutter, were already studied in sufficient detail (see, e.g., the surveys [34-36]) at the time we began our investigation. It is clear qualitatively that the output power of such a laser will increase as the initial transmission of the filter decreases, since this leads to accumulation of a large number of particles on the upper laser level. In this connection, the use of filters will minimal initial transmission is advantageous from the viewpoint of obtaining maximal emitted power. However, in practically all cases studied, lasing could be achieved only at high initial transmissions of the filter (of the order of several percent) due to the constraints imposed by the limiting value of the gain of the active medium. It has been pointed out in the literature [37, 38] that the use of ruby at low temperatures as the active medium is of promise in this respect. As is seen from the graph in Fig. 2, values of K_{thr} of the order of 0.5 cm^{-1} are attainable. Nevertheless, extremely little work has been done on the experimental investigation of Q-switching in ruby lasers at low temperatures [39-41], and in only one study [40] was a bleachable filter used for the Q-switching. The parameters of the experimental setup used in [40] were very far from optimal, and the results obtained were not of much help in subsequent studies in this area.

We briefly consider the main types of lasing regimes in lasers with bleachable filter: The giant pulse and the self-mode-locking regimes. The giant pulse regime is the simpler from the point of view of the methods required to achieve it and in terms of its theoretical description. The giant pulse regime can already be described by arguing on the level of the balance equations of type (1.1), from which the term describing the change of inversion due to the pumping is omitted. This argument was carried out in [37], where the time behavior and emitted energy were calculated as a function of the drop in Q, assuming instantaneous Q-switching and a uniform distribution of inversion and field density over the volume of the resonator.

We note that the paper [37] was published before the appearance of the first papers on lasing by lasers employing bleachable filters as passive shutters [42, 43], and the fundamental purpose of [37] was to describe processes which take place during active Q-switching. Nevertheless, in many cases [37] gives quite a good description of the character of the lasing kinetics in lasers with a passive shutter, and even permits a direct quantitative comparison with experimental results, although when the duration of a giant pulse and the emitted energy are calculated on the basis of the data in [37] one in general obtains values for the duration and energy which are too low and too high, respectively.

A theoretical investigation of the lasing process with allowance for the finite time required to bleach the filter was carried out in [44] and, in more detail, [45]. The results of these papers provide a justification for supposing that bleaching of the filter may be assumed instantaneous if $\sigma_{b.f.}/\sigma \geq 100$ ($\sigma_{b.f.}$ is the absorption cross section for the material composing the bleachable filter). This condition certainly holds for polymethine dyes used in ruby lasers with a passive shutter.

In [46] the parameters of a giant pulse were investigated taking into account the finite relaxation time $\tau_{b.f.}$ of the dye molecules, and a condition was obtained under which the effect of relaxation is negligibly small,

$$2\sigma_{b.f.}\Phi_{max}\tau_{b.f.}\frac{n_{thr}}{n'_{thr}} > 100, \qquad (2.1)$$

where Φ_{max} is the maximal photon flux density, n_{thr}/n'_{thr} the ratio of the threshold inversions before and after bleaching of the filter. This condition is in general satisfied in a ruby laser.

In all the theoretical papers mentioned above, no allowance was made in the analysis for spatial inhomogeneities in the distribution of the emission field or the inversion. Clearly, the theoretical values of the pulse durations obtained in such a treatment should be too low. Indeed, inhomogeneity of the transverse distribution of inversion leads to a situation in which the inversion at different points of a cross section of the active element burns out at different times, and the experimentally observable form of a giant pulse is a convolution of the giant pulses generated at different points of the rod cross section. This phenomenon was observed experimentally in [47] in ruby at low temperature. The process of transverse development of the field of a giant pulse in an active medium with an inhomogeneous inversion distribution was studied theoretically in [48] by numerical solution on a computer. Nonuniformity of the lasing field in the longitudinal direction, which is most important for an active medium with a large gain, was taken into account in [38, 49]. It is shown there that if the duration of a giant pulse is of the order of 2L/c, then in a high-Q resonator amplitude modulation of emission occurs with axial period 2L/c even if interaction of the axial modes is ignored.

In [50] spatial coherence of emission in the giant pulse regime was investigated, and it is shown that two conditions must be satisfied in order for the emission to be spatially coherent. The first condition coincides with (1.6) with the inequality reversed, and is the condition for one transverse mode to be generated. The second condition can be expressed as

$$|\ln \eta| < \frac{\pi L \lambda}{4a^2},\qquad\qquad(2.2)$$

where η is the initial transmission of the filter and a is the transverse dimension of the mode. This condition means that distortion of the mode field due to burnout of inversion does not occur during the lasing process. An approximate calculation shows that the emission will be spatially coherent only if $\eta > 50\%$.

The self-mode-locking regime is qualitatively different from the giant pulse regime and was first observed experimentally in [51-53]. This regime cannot be described in the language of balance equations since the latter are obtained when interaction effects among the modes are neglected. In this regime, laser emission consists of a regular train of pulses separated by the axial period. In the language of modes, this means that the phases of all the lasing axial modes are synchronized. Exactly the same situation arises in active mode-locking which is achieved at resonance (period 2L/c) modulation of losses [54]. In contrast to active mode-locking, self-mode-locking develops spontaneously when there is suitable selection of transverse modes.

The theory of self-mode-locking in lasers with bleachable filters is discussed most completely in [35, 36]. The description given in these papers is based on considering the time evolution of the emission intensity profile, as a result of which separation of the ultrashort pulses from the noise background occurs. According to [35, 36] the development of an ultrashort pulse can be broken into several stages, in each of which the interaction of radiation with the material of the bleachable filter takes place differently.

The development of lasing can be pictured arbitrarily as follows. In the linear stage when the intensity of radiation is not sufficient to bleach the filter, linear amplification of the quasiperiodic fluctuation type occurs, accompanied by a narrowing of the emission spectrum (this phenomenon is analogous to the narrowing of the spectrum which occurs in ordinary free-running lasing [55]). During the nonlinear stage of lasing, the profile of the lasing field is transformed as the filter becomes bleached, so that points of the profile where the intensity is maximal are increased preponderantly. As a result of repeated passes of radiation through the bleachable filter, each individual spike in the intensity becomes narrowed in time, and simultaneously the most intense spikes become more pronounced. Complete mode locking corresponds to the case when practically all the energy of laser emission over an axial period is concentrated in a single spike.

In this approach it becomes obvious that self-mode-locking is never completely reproducible. In each laser with a passive shutter there exists only a certain probability that complete self-mode-locking will be achieved, and its value depends on the properties of the bleachable absorber and the laser parameters. The notion of complete self-mode-locking now becomes to some extent arbitrary. For example, [35] takes the ratio of energy contained in the most intense pulse to the energy contained in all the remaining pulses over an axial period as the criterion for the degree of mode-locking. Naturally, the calculated value of the probability of obtaining self-mode-locking depends on the value chosen for this ratio.

The validity of the qualitative picture of lasing proposed in [35, 36] is supported by the experimental results obtained in [56], where a photoelectron detector was used to investigate the process of development of lasing of ultrashort pulses in a neodymium-glass laser, starting with irregular noise and ending with a periodic sequence of ultrashort pulses.

At the same time, it should be remembered that due to a series of assumptions made in [35, 36], many characteristics of real lasers, in particular the lasing spectrum and the shape of individual pulses, cannot be calculated directly on the basis of the results of these papers. An extremely important difference between real and idealized lasers is the finite relaxation time of the bleachable absorber. Clearly, the minimal duration of an ultrashort pulse formed during the nonlinear stage of lasing cannot be less than this relaxation time. A rigorous allowance for this effect is extremely difficult in the framework of the approach used in [35, 36]. However, it is perfectly possible to give a qualitative idea of the situation using simple physical arguments given in [35, 36]. It is natural to suppose that the inertia of the filter plays a role when $\tau_{b.f.}$ is comparable to the reciprocal width of the spectrum $\Delta\nu$ formed toward the start of the nonlinear stage of lasing,

$$(\Delta\nu)^{-1} \lesssim \tau_{b.f.} \tag{2.3}$$

If $\tau_{b.f.} > (\Delta\nu)^{-1}$, then the bleachable filter separates from the noise a group of pulses whose envelope has a duration of the order of $\tau_{b.f.}$. The characteristic duration of the individual pulses is $(\Delta\nu)^{-1}$. If $\tau_{b.f.} \simeq (\Delta\nu)^{-1}$, a smooth pulse of duration $\simeq \tau_{b.f.}$ is selected. These phenomena were observed in ruby lasers at room temperature in [57] when $(\Delta\nu)^{-1}$ was varied by introducing discriminating elements into the resonator.

The effects of the finite relaxation time of the bleachable filter are strongest in Nd-glass lasers, since the characteristic values of $\tau_{b.f.}$ for filters at the wavelength 1.06 μm lie in the range 7-11 psec [58], which is much larger than the reciprocal width $(\Delta\nu)^{-1}$ of the spectrum. In the cryptocyanine and DDI dyes used to obtain self-mode-locking in ruby lasers, the value of $\tau_{b.f.}$ is 22 and 14 psec, respectively, in acetone solution [59], which is comparable to $(\Delta\nu)^{-1}$. These values are in good agreement with the smallest pulse durations presently attainable in the self-mode-locking regime. When DDI is used, pulses of duration 15 psec [60] are obtained, while pulses of duration 25-30 psec [61] are obtained with cryptocyanine.

Thus the finiteness of $\tau_{b.f.}$ is one of the main factors limiting the maximal attainable lasing power in ruby lasers with mode locking (in neodymium-glass lasers, a decisive role is played by self-modulation [62] and self-focusing [63] which arise because of nonlinearity of the index of refraction at times when a high field density is attained in the resonator). In this context, the problem of designing a passive shutter having the smallest possible relaxation time is of great current interest. One of the possible ways to decrease $\tau_{b.f.}$ for the dyes already known is proposed in [64]. A significant shortening of the pulses in mode-locking regimes was obtained there by using a very thin (to 3 μm) cell in which the dye was dissolved, so that a corresponding increase in the dye concentration was achieved.

In [35] certain other factors are discussed which affect the final field profile, in particular, saturation of the gain, dispersion of the index of refraction, frequency self-modulation of emission, the finite thickness of the cell containing the dye, and nonlinear effects (self-focusing, multiphoton absorption, etc.).

§2. Generation of a Giant Pulse in a Ruby Laser

at Low Temperature

In the investigation of the lasing characteristics in the giant pulse regime, the usual laser setup with a passive shutter was used. Roughly polished ruby rods of length 230 and diameter

14 mm (crystal A) and length 120, diameter 12 mm (crystal B) were used as the active elements. The ends of the rubies were cut at the Brewster angle. The resonator was formed from two flat mirrors with reflection coefficients 30 and 100%. The optical length of the resonator was 80 cm. A cell of thickness 5 mm containing a solution of cryptocyanine in ethanol placed between the ruby and the opaque mirror served as the filter. The initial transmission of the filter varied from 25 to 0.03%.

The characteristics of pulse generation were measured as far as possible under thresh-old conditions. However, at high initial transmissions (greater than 1% for crystal A and greater than 10% for crystal B) it was not possible to obtain single peaks. Rather, several pulses were excited with an interval of 40-80 μsec. If the initial transmission became less than 0.03%, self-excitation of lasing was observed for crystal A due to parasitic feedback and a transition to a "hard" regime of bleaching of the filter occurred.

The results obtained for crystal A are mainly discussed below, since the main charac-teristics of giant pulses for both crystals were nearly the same for identical filter densities. The only exception is an interesting effect which could only be observed for crystal B. This effect, called self-modulation of emission [49], consists of modulation of the amplitude with axial period on the trailing front of a giant pulse. It will be considered separately at the end of this section.

Figure 14 shows examples of oscillograms of giant pulses for different initial values of the transmission of the bleachable filter. The oscillograms were obtained using an FÉK-09 photoelement whose signal was fed into an I2-7 oscillograph. The time constant of the record-ing system was not more than 1 nsec. The pulse build-up time was in all cases less than the decay time, which is characteristic for the case of instantaneous Q-switching [44]. The oscil-lograms obtained were used to measure the pulse duration t_{pulse} for filters with η = 25; 13; 10; 5; 1; 0.3; 0.2; 0.1; 0.03%. The experimental values of t_{pulse} as a function of η are shown in Fig. 15 (points).

Measurements were also performed with a resonator in which a diaphragm was placed between the ruby and the filter. The diaphragms were of diameters 3.2, 2.5, and 1.5 mm. Figure 15 shows the results of the measurements. The decrease in the pulse duration was achieved mainly due to shortening of the trailing front (its duration remained as before greater than the duration of the leading front).

Figure 15 also shows a theoretical graph of t_{pulse} as a function of η, calculated on the basis of the results of [37]. It seems to us that a comparison of the experimental values of t_{pulse} with the values calculated in [37] is justified, since in the present case $\sigma_{b.f.}/\sigma \simeq 10^3$ and

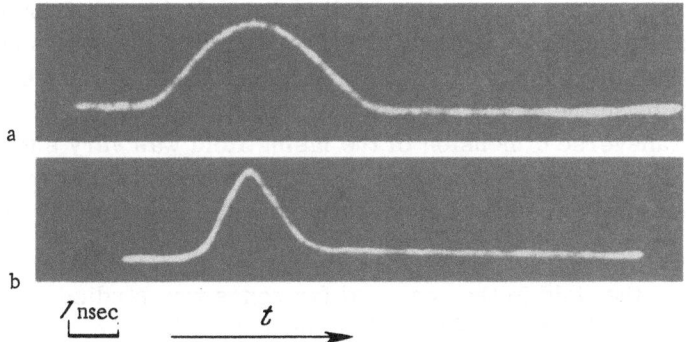

Fig. 14. Oscillograms of giant pulses a) η = 1.0%;
b) η = 0.03%.

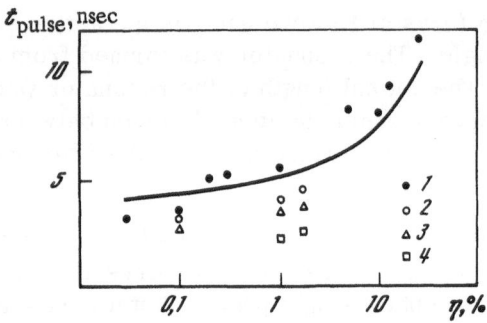

Fig. 15. Dependence of the duration of giant pulses t_{pulse} on the initial transmission of the filter η. Solid curve values of t_{pulse} calculated from data of [37]; points, experimental: 1) in a resonator without diaphragm; 2) with diaphragm, 3.5 mm; 3) with diaphragm, 2.5 mm; 4) with diaphragm, 1.5 mm.

bleaching of the filter may be assumed instantaneous. In addition, since condition (2.1) is satisfied, relaxation of excited molecules in the bleachable absorber can be neglected. As is seen from the figure, the agreement between the calculated and experimental values of t_{pulse} is fairly good. The observed deviations can apparently be explained by inaccuracy in measuring the transmission of the filters (the error for dense filters can reach 50%) and by effects related to the nonuniform distribution of inversion over the volume of the resonator in the longitudinal and transverse directions. In particular, an inhomogeneous transverse distribution of inversion causes a certain lengthening of the pulse due to the fact that as the transverse lasing field develops, different transverse modes commence generation nonsimultaneously. The introduction of a diaphragm into the resonator should limit the transverse development and shorten the pulse, and this is observed in our experiments. In addition, in lasers with long diaphragms when the Fresnel number is small, transverse development of the field also occurs as a result of divergence of the emitted radiation, as follows. After the start of lasing in the region of maximum inversion and after bleaching of the filter by the main lasing field, the region of transluscence will expand due to divergence of the emission. The time for transverse development of the pulse can be easily estimated if the angular divergence of emission and the dimension of the near zone are known. In our case, these values were determined from integral photographs of the field distribution and were $8 \cdot 10^{-3}$ and 0.5 cm, respectively. This gives a value of the order of 2 nsec for the time for transverse development. As is seen from Fig. 15, the introduction of a diaphragm into the resonator, which obstructs transverse development of the pulse, causes a shortening of the pulse by approximately the same amount.

An analogous transverse development discovered at room temperature in [47] occurs at larger times (10-20 nsec). This difference is due to the fact that at room temperature the primary divergence is smaller, of the order of $5 \cdot 10^{-4}$. The increase of the divergence in our case probably results from significant inhomogeneities in the distribution of the gain, as well as from scattering by inhomogeneities in the ruby crystal (the optical quality of crystal A was very low). Clearly, for the same reason no amplitude modulation of emission with period $2L/c$ was observed in the experiment with crystal A, for as shown in [38, 49], such modulation should arise as a result of longitudinal inhomogeneities in the distribution of inversion and the density of the field when the gains in the active medium are high. Even when a diaphragm was placed in the resonator, the transverse dimension of the lasing field was very substantial due to the large divergence of emission, and the fact that lasing occurred at different times at different points of the cross section of the crystal apparently resulted in smearing out of the amplitude modulation.

The distribution of the field in the near and far zones was studied in terms of integral photographs. When the initial density of the filter increased, the size of the region of lasing and the angular divergence of emission did not change significantly, but the distribution of the field became more uniform, which indicated a strong distortion of the transverse modes in the

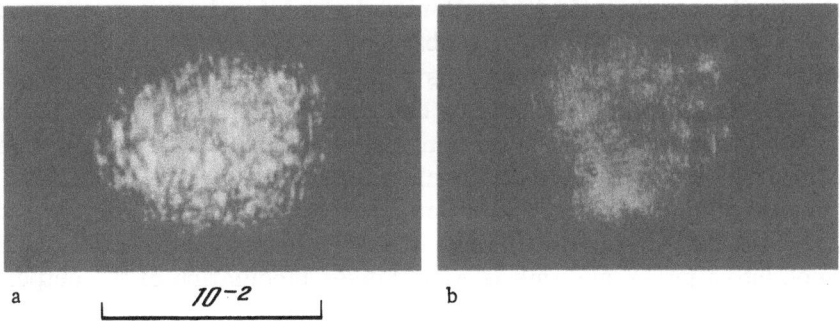

Fig. 16. Integral photographs of the distribution of the field of a
giant pulse in the far zone: a) $\eta = 10\%$; b) $\eta = 0.5\%$.

lasing process. This result is in agreement with the ideas developed in [50, 65]. Examples of integral photographs of the field in the far zone are shown in Fig. 16.

The pulse energy was measured using an IZhK-1 device. The number of pulses was monitored by an SFR-2M photorecorder in the slit-scanning mode. In the case of several pulses, the average energy per pulse was calculated. The experimental values of the energy W obtained are given below.

η, %	10	4	1.8	1	0.3	0.06	0.03
W, J	0.10	0.12	0.16	0.23	0.36	0.34	0.36

The maximal emitted power calculated from the data on the duration and energy of the pulse is of the order of 100 MW.

The theoretical estimate of W made in accordance with the data of [37] gives values two to three times greater than the experimental values. This discrepancy is apparently caused by the fact that in practice it is very difficult to determine both the effective lasing volume and the average value of the inversion in this volume. In any case, it is obvious that the effective surface area of the lasing volume should be less than the area determined from the integral photographs of the near zone. In addition, it was assumed in the calculation that the filter is completely bleached. In actuality, a filter in the bleached state possesses a residual absorption, the value of which depends on the initial transmission of the filter [66].

The width Δ_{las} of the lasing spectrum was measured by a Fabry−Perot interferometer with base 30 mm.

The experimental results as well as the theoretical values calculated from Eq. (1.12) are given below.

η, %	10	1.8	0.5	0.05
$\Delta_{las\ exp}$, cm^{-1}	0.060	0.047	0.045	0.043
$\Delta_{las\ theor}$, cm^{-1}	0.052	0.038	0.038	0.040

The value of Δ_{las} appearing in Eq. (1.12) was determined with the aid of a Fabry−Perot interferometer and was equal to 2.3 cm^{-1}. The discrepancy between the theoretical and experimental values of Δ_{las} lies within the limits of experimental error.

We now stop to discuss the effect of self-modulation of radiation from a giant pulse which we observed for crystal B when a diaphragm obstructing transverse development of the lasing field and eliminating lasing by high-order transverse modes was introduced into the resonator. As noted in [38, 67], self-modulation should be observed when the Q-switching time and the characteristic time of variation of the inversion are comparable with the axial period of the resonator. The mechanism giving rise to modulation of the pulses is as follows: when Q-switching is rapid, longitudinal inhomogeneity of the field density (a step-jump) forms in the resonator and propagates along it. As a result of saturation of inversion, the step-jump becomes transformed into a pulse, and this is what causes modulation of the output emission.

Amplitude modulation of a giant pulse was first observed by us in lasers with resonator with the parameters described above. However, the presence of modulation in this case could be caused not only by the mechanism described above, but also by the fluctuation mechanism [35, 36] responsible for self-mode-locking. For this reason a special investigation was undertaken to try to obtain the self-modulation effect in a "pure" form. The resonator parameters were changed somewhat. The output mirror with reflection coefficient 30% was replaced by a mirror with reflection coefficient 70%. The bleachable filter was placed flush against the output window.

In order to study the effect of self-modulation in pure form, it was necessary to assure a homogeneous distribution of the emission field in the longitudinal direction at the moment the passive shutter was opened, in order that inhomogeneity of the field responsible for amplitude modulation should develop only as a result of rapid saturation of amplification in the active medium. In spectral language, this corresponds to the need to assure single-mode lasing at the instant the shutter is opened. To this end, a selector was introduced into the resonator which consisted of a glass etalon of thickness 3 mm with reflecting coverings (reflection coefficient $\simeq 30\%$). An investigation of the lasing spectrum using a Fabry–Perot interferometer with base 50 mm (which provided for resolution of neighboring axial modes) showed that the selector permits obtaining a single-mode regime in cases when the initial transmission of the bleachable filter is not less than 10%. When denser filters are used, several modes commence lasing immediately after the shutter is opened, which makes it impossible to observe the self-modulation effect in pure form.

Oscillograms of the lasing pulses are shown in Fig. 17. It is seen that modulation develops at the moment when maximal density of emission is attained, at which time a rapid drop in

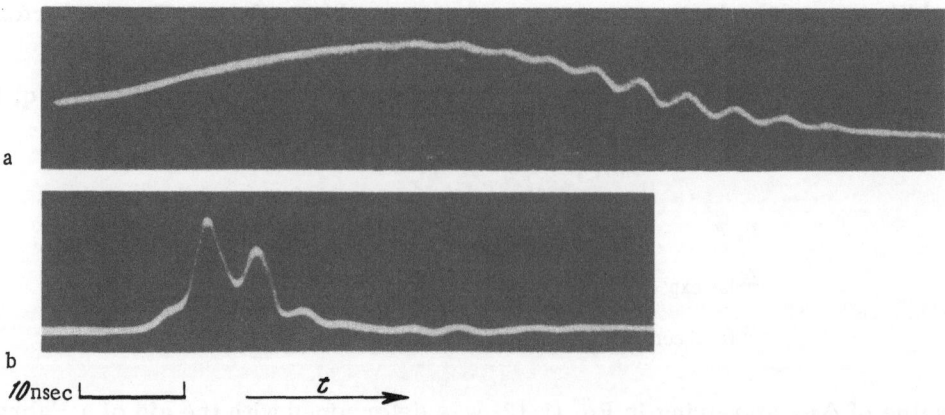

Fig. 17. Oscillograms of laser emission. Initial transmission of filter:
a) 30%; b) 10%.

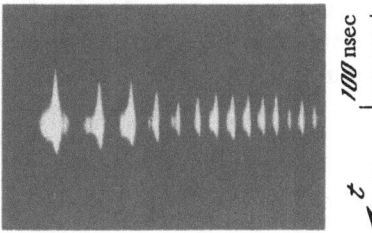

Fig. 18. Slit scan of emission spectrum. Initial transmission of filter, 30%.

the inversion in the active medium occurs. The extent of modulation increases as the initial density of the bleachable filter increases. Figure 18 shows a time scan of the lasing interference pattern obtained using the SFR-2M. The lasing in the first instants after bleaching of the filter is single-mode. The appearance in the spectrum of lateral bands corresponding to development of amplitude modulation of emission with the axial period is observed at the time of saturation of inversion in the active material. In our view, this type of time dependence of the emission intensity and spectrum is experimental proof of the existence of the self-modulation effect. Here the only role of the bleachable filter is in Q-switching.

The qualitative dependence of the extent of modulation on the Q-switching time was studied, the latter depending on the ratio of the absorption cross sections of the bleachable filter and the active medium. For this investigation a telescope was introduced into the resonator to increase the cross-sectional area of the beam before it passed through the bleachable filter. This in fact corresponds to a decrease of the effective absorption of the material of the bleachable filter. It was established that the extent of modulation decreases with decreasing cross section. There was practically no modulation when the quantity $\sigma/\sigma_{b.f.} \tau_{b.f.}$ characterizing the time required to open the bleachable filter became approximately equal to $2 \cdot 10^8$ sec^{-1} (10^7 sec^{-1} without telescope).

Within the limits of experimental accuracy, no correlation was discovered between the extent of modulation and the length of the resonator, nor between the extent of modulation and the position of the active element in the resonator.

§3. Self-Mode-Locking in Ruby Lasers at Low Temperature

The ruby crystals described in the preceding section were used in the self-mode-locking experiment. The resonator geometry remained as before in most respects. The reflection coefficients of the mirrors were 70 and 100%. The optical length of the resonator was 65 cm and the cell of thickness 5 mm was replaced by a cell 1 mm thick in contact with the opaque mirror. Two diaphragms of diameters 2.5 and 1.8 mm were placed in the resonator in order to suppress extra-axial modes.

A mixture of nitrobenzene and toluene in the proportion 1:4 was used as the solvent for cryptocyanine. This choice of solvent was made because of the need to make the indices of refraction of the dye solution and the glass support (serving as the front wall of the cell) equal. If these indices are not equal it is possible for generation to occur due to reflection from the glass-filter interface. In this case a transition will occur to a "hard" bleaching regime, in which the filter is bleached by laser emission originating in a region exterior to the filter. At the same time, in order for self-mode-locking to occur it is necessary that there be a "soft" regime of bleaching of the filter, i.e., the free-lasing emission which bleaches the filter should develop through the filter. In our case the index of refraction of the solvent was approximately equal to the index of refraction of K-8 glass.

Self-mode-locking was studied for filters with initial transmissions from 1 to 50%. Complete self-mode-locking occurred in 90-95% of the cases, which corresponds to the probability P of locking calculated from the equation [35].

$$e^{-\frac{1}{p}[\ln m + \ln M]\ln m} < P < e^{-\frac{1}{p}\ln M \ln m} \qquad (2.4)$$

Here $p = (\beta - \gamma_l)/(\beta - \gamma_l - \gamma_{nl})$ (β is the amplification in the active medium toward the end of the linear stage of lasing, $\beta = Kc$ are the linear losses in the resonator; γ_{nl} the nonlinear losses in the resonator which characterize the initial transmission of the bleachable filter); m, the number of axial modes which decay over a linewidth $\Delta\nu$; M, the ratio of the energy contained in the most intense pulse over an axial period to the energy contained in the other pulses (a parameter characterizing the "quality" of mode-locking). Values of the parameters p = 100, m = 10, M = 10, corresponding to the parameters of our laser were taken in the calculation.

Figure 19 shows oscillograms of laser emission for crystal A in the self-mode-locking regime. An additional spike is seen in Fig. 19 in addition to a self-mode-locking pulse of half-width 0.5-0.7 nsec. As will be shown in the next section, this spike does not represent the true time behavior of emission and is probably caused by reflections in the channel of the recorder. The shape of the envelope of the wavetrain for self-mode-locking is close to the shape of a smooth giant pulse, the generation of which occurred in the absence of mode-locking. Its duration also decreased with increasing initial filter density. However, for equal initial filter densities, the duration of the envelope was approximately 1.5 times greater than for the corresponding giant pulse.

The case of mode-locking at small initial transmissions of the filter is of particular interest, since here the duration of the corresponding giant pulse was less than the trace time of pulses under mode-locking conditions, and it became possible to isolate a single ultrashort pulse. Selection of an isolated ultrashort pulse is at present done mainly by two methods — using an optical shutter [68] or using a burn-through mirror [69]. The advantage of the first method is that it is possible to select any group of pulses on any segment of the wavetrain. However, this method requires auxiliary equipment which leads inevitably to a complication of the experimental setup. The second method is the simplest, but in this case pulse selection occurs in the build-up part of the wavetrain, and therefore the intensity of the selected pulse is not maximal. By decreasing the initial transmission of the filter to 1%, we succeeded

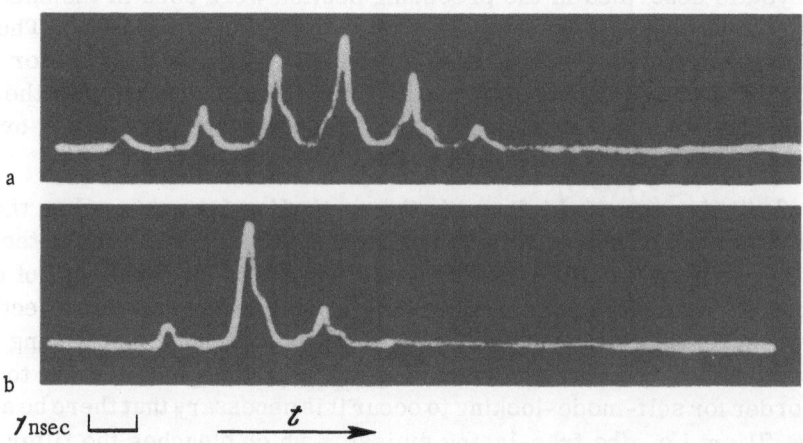

Fig. 19. Oscillograms of laser emission in the self-mode-locking
regime: a) $\eta = 8\%$; b) $\eta = 1\%$.

in our experiments in obtaining selection of an isolated pulse with probability 50% (Fig. 19a). In these cases the amplitude of the largest pulse was more than twice as large as the amplitude of the other pulses.

Also investigated was the possibility of obtaining self-mode-locking when denser filters were used. It turned out that the probability of mode-locking drops abruptly for filters with initial transmission less than 0.5%. This is evidently related to the development of free lasing due to reflection at the glass-filter boundary as a result of differences in the indices of refraction and transition to a "hard" regime of filter bleaching. The values of the energy of a wavetrain measured using an IZhK-1 device for filters with initial transmissions 8, 3.5, and 1% are given below.

η, %	8	3.5	1
W, J	0.02	0.03	0.08

Here as in the preceding section, the number of ultrashort pulse trains was monitored with the aid of an SFR-2M photorecorder, and when several trains were present the average energy for a single train was calculated.

An investigation of the width of the spectrum using a Fabry−Perot interferometer showed that the widths differ considerably for crystals A and B. For crystal A, the average width of the lasing spectrum under self-mode-locking conditions did not depend on the initial filter density and was approximately 0.08 cm^{-1}, which corresponds to a limiting pulse duration of 0.3-0.4 nsec and maximal emission intensity of 7 GW/cm^2 for a filter with initial transmission of 1%. Some increase in Δ_{las} compared with the smooth giant pulse regime occurred due to shortening of the pulse duration during the nonlinear stage of lasing as the filter was bleached. However, the value of Δ_{las} calculated from [35] exceeded the experimental value by two to three times. This difference could be due to the fact that shortening of the pulse during the nonlinear stage of lasing is hindered by strong scattering of radiation in a ruby crystal of low quality. For crystal B the average width of the lasing spectrum for filters of initial transmission 20% and above was approximately 0.2 cm^{-1}, which corresponds roughly to the theoretical value. When the initial transmission of the filter was decreased, the lasing spectrum became wider, and for $\eta = 3\%$ the average spectral width reached 0.4 cm^{-1}. We suggest an interpretation of this result in the next section.

§4. Effect of the Relaxation Time of the Dye on the Pulse Shape during Self-Mode-Locking in Ruby Lasers at Low Temperature

As already remarked in Sec. 1 of this chapter, it is shown in [57] that one of the important factors determining the duration of ultrashort pulses in the self-mode-locking regime is the relaxation time $\tau_{b.f.}$ of the bleachable filter. By means of reducing $\tau_{b.f.}$, a considerable increase in the width of the lasing spectrum and shortening of individual ultrashort pulses was obtained in [64] for self-mode-locking in a neodymium laser. In that paper the reduction in $\tau_{b.f.}$ was achieved by increasing the concentration of dye and was caused by an efficient transition of the dye molecules from the upper level due to superradiance [70].

We used a somewhat different approach, wherein τ_{las} was controlled by changing the composition of the solvent. Quite reliable methods for measuring τ_{las} have recently been developed [59, 71]. The results of [59, 71] indicate that τ_{las} and hence the duration of individual ultrashort pulses can be regulated over wide limits by altering the composition of the solvent.

The laser setup used in this experiment was the same as in the preceding experiment. Crystal B was used as the active element. A solution of cryptocyanine in a mixture of nitro-

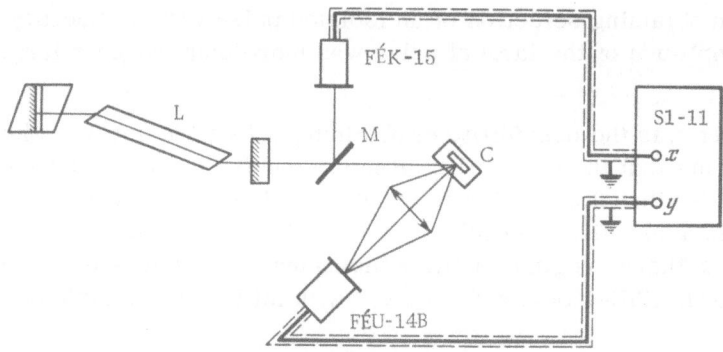

Fig. 20. Diagram of experimental setup for measuring
the relative quantum yield from luminescence of cryptocyanine. L, laser; M, semitransparent mirror; C, cell
containing cryptocyanine solution.

benzene and toluene was used as the bleachable filter. In order to vary $\tau_{b.f.}$, the quenching
agents carbon tetrachloride and acetone were added to the solution.

In order to determine the efficiency of quenching of the excited level of cryptocyanine,
measurements were made of the relative quantum yield B from luminescence of cryptocyanine
as a function of the concentration C of quenching agent. Figure 20 shows the setup for measuring B. Radiation from a ruby laser in the free-running mode was used to excite luminescence
of cryptocyanine, a solution of which was poured into a cell of thickness 5 mm. Part of the
radiation (about 10%) was diverted onto an FÉK-15 photoelement, the signal from which passed
onto the horizontal deflecting plates of an S1-11 oscilloscope. The cryptocyanine luminescence
was recorded by an FÉU-14B photomultiplier, the signal from which was fed into the vertical
deflecting plates of the oscilloscope. A time delay line ($\simeq$ 600 nsec) was put into the circuit of
the FÉK in order to compensate the phase difference in the FÉK and FÉU signals, which arose
because of the large time of transit of electrons in the FÉU. The laser emission scattered toward the FÉU was suppressed by KS-19 light filters having equal thickness so chosen that
when an interference light filter was placed in front of the FÉU at the wavelength 693.4 nm (the
wavelength of ruby laser emission at 90°K) there was no signal on the vertical deflecting plates
of the oscillograph. Thus the trace on the oscillograph screen was a graph of the dependence of
the intensity of luminescence of cryptocyanine as a function of the intensity of the exciting radiation. In the absence of saturation of absorption, the graph was a straight line, the slope of which
is proportional to the quantum yield of luminescence. The value of the relative quantum yield of
cryptocyanine luminescence in different solutions (relative to a solution in nitrobenzene) could
be determined from the ratio of the slopes on the oscillograms. The optical thickness of the
solution was chosen so that total absorption of the exciting radiation occurred. The concentration of cryptocyanine in the solution was approximately the same as for the bleachable filter
experiments investigating self-mode-locking.

The experimental results are presented in Table 1 together with the corresponding absolute values of $\tau_{b.f.}$, which were estimated by using the value of $\tau_{b.f.}$ for a solution of cryptocyanine in acetone (according to [59], $\tau_{b.f.} = 22 \pm 4$ psec).

We note that the experimental values of $\tau_{b.f.}$ for pure nitrobenzene appear 30-40% lower.
This is connected with the fact that the center of the luminescence line of cryptocyanine in
nitrobenzene is shifted by 200-300 Å toward the infrared as compared to the center of the
luminescence line of cryptocyanine in acetone [72]. The sensitivity of the photoelement of the

TABLE 1

	C, %				
	0	20	50	80	100
B					
acetone	1	0.82	0.64	0.51	0.41
CCl$_4$	1	0.84	0.76	0.69	0.55
$\tau_{b.f}$, psec					
acetone	54	44	34	27	22
CCl$_4$	54	45	41	37	30

FÉU-14B in this range of wavelengths depends quite strongly on the wavelength of the radiation registered.

The lasing spectra under self-mode-locking conditions were recorded with the aid of a Fabry–Perot IT-51-39 etalon with bases 3 and 1 mm. In order to avoid overlapping of the spectra during generation of several trains of pulses, as occurs for small initial filter transmissions (cf. Sec. 2), the image of the spectrum was scanned by an SFR-2M photorecorder. The spectrum was studied only for the first train of pulses, since in succeeding trains the quality of mode-locking was poorer in general (in some cases it was altogether absent, and generation of smooth giant pulse occurred). Figure 21 shows a graph of the dependence of the width of the lasing spectrum on the concentration of acetone C in solution for the two values 3 and 50% of the initial filter transmission, η. Each point on the graph was obtained by averaging over ten interference patterns. The length of the vertical segments correspond to the spread in the values obtained in the individual measurements. The relative fluctuations of the spectral width Δ_{las} in this experiment were much smaller than in [64], where Δ_{las} varied from 2 to 60 cm^{-1} from flare to flare. This is due to the fact that the ratio of $(\Delta\nu)^{-1}$ to $\tau_{b.f.}$ was essentially different in these experiments. Indeed, in [64] $\tau_{b.f.} > (\Delta\nu)^{-1}$, and therefore the bleachable filter selected from the initial noise a group of ultrashort pulses, the spectrum of which fluctuated from flare to flare. In our case, $\tau_{b.f.} > (\Delta\nu)^{-1}$ and selection of an isolated smooth pulse occurred, the fluctuations in the duration of individual pulses being smoothed out owing to the finite relaxation time of the dye.

At acetone concentrations of 50% and greater, the width of the spectrum is almost independent of C. This means that at such concentrations the filter may be assumed to have no inertia. Indeed, for C ≥ 50% the width of the spectrum corresponds to the theoretical value calculated by the formula [35]

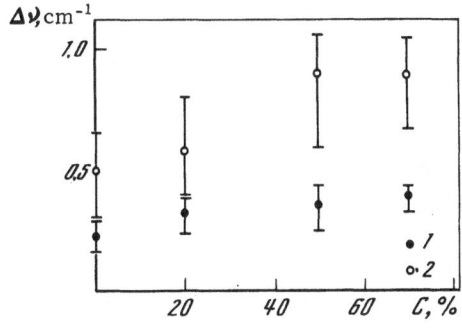

Fig. 21. Dependence of the width of the lasing spectrum on the concentration of acetone in solution. 1) η = 50%; 2) η = 3%.

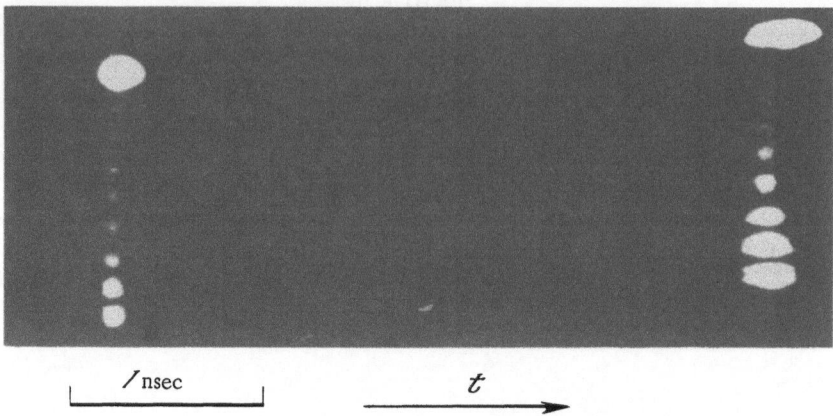

Fig. 22. Chronogram of laser emission ($\eta = 3\%$, C = 50%).

$$\Delta_{las} = \frac{\Delta_l}{\sqrt{Z}} \sqrt{\frac{\gamma_{nl}}{\gamma_l + \gamma_{nl}}}, \qquad\qquad (2.5)$$

where $Z = \beta(t_1 - t_{thr})^2$ (β is the rate of build-up of amplification at the time threshold is reached; $t_1 - t_{thr}$ is the duration of the linear stage of generation).

When carbon tetrachloride was used as the quenching agent, completely analogous results were obtained. The experimental average values for Δ_{las} in this case differed from those shown in Fig. 21 by less than the fluctuations of Δ_{las} from flare to flare.

In order to check the correspondence of the width of the spectrum and the duration of an individual ultrashort pulse, the shape of the pulses was studied using an image converter tube with a time scan. Figure 22 shows chronograms of laser emission for filters with $\eta = 3\%$ and C = 50%. The vertical lines on the chronograms correspond to the images of individual components of a step attenuator placed at the output slit of the image converter. The distance between pulses corresponds to the time for a double passage across the resonator, equal to 3.5 nsec. It is seen that only one pulse of duration not more than 300 nsec is present on an axial period. Consequently, the additional spikes which were visible on the oscillograms of the laser emission under mode-locking conditions (Fig. 19) were the result of reflections in the channel of the recorder. The time behavior of the emission in the individual pulses for filters with different η and C, calculated from microphotograms of the photographs obtained using the image converter, is shown in Fig. 23.

It is seen that the main effect of the finiteness of the relaxation time of cryptocyanine is to stretch out the trailing edge of the pulse, while the shape of the leading edge is practically unchanged. When a quenching agent is added to the solution, the pulse acquired a more symmetric shape. The fact that the pulse selected by the bleachable filter is smooth was a consequence of the relation $(\Delta\nu)^{-1} \geq \tau_{b.f.}$ which held in our case. In terms of the intensity distributions obtained, it was possible to estimate the duration of individual pulses bearing in mind the fact that the shape of the pulses observed experimentally is the convolution of their actual shape with the apparatus function, the width of the latter being 30 psec. The average values of the pulse durations, taking into account the width of the apparatus function, for a filter with initial transmission 3% were 35 and 80 psec, respectively, for a 50% concentration of quenching agent and without quenching agent, the corresponding values being 180 and 260 psec for a filter with initial transmission of 50%. The results of the measurement provided grounds for assuming that the

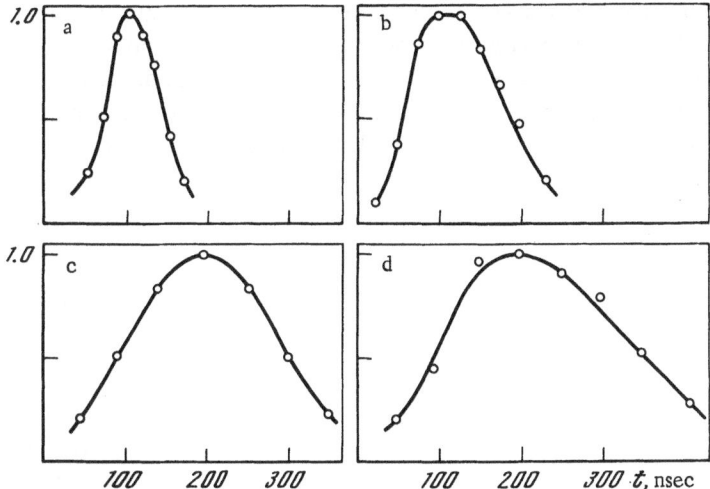

Fig. 23. Time behavior of emission in an individual pulse.
a) $\eta = 3\%$, C = 50%; b) $\eta = 3\%$, C = 0; c) $\eta = 50\%$, C = 50%;
d) $\eta = 50\%$, C = 0.

duration of individual pulses corresponds to the width of the spectra given in Fig. 21, i.e.,
$\Delta_{las} c\tau_{pulse} \simeq 1$.

The decrease in the pulse duration with increasing initial transmission of the filter is an interesting effect. The experimental values of Δ_{las} are 3-4 times greater than the theoretical values calculated from Eq. (2.5). In certain cases the measured values of Δ_{las} approached the value of Δ_l, which for the ruby sample used was 1.7 cm^{-1}. In our view, two facts must be considered in explaining this result. First, as is seen from the lasing oscillograms, strong saturation of amplification occurs. The inversion is destroyed practically after a single passage across the resonator. Second, when $\eta = 3\%$ the pulse duration becomes smaller than the relaxation time T_2 in ruby (of the order of 100 psec at 90°K [14, 15]). Accordingly, it can be proposed that the shortening of the pulses is due to two reasons related not to the action of the bleachable filter but to the properties of the interaction of radiation with the amplifying medium.

After the nonlinear stage of lasing has ended, the duration of the selected pulses is evidently the same for all densities of the bleachable filter. However, after complete bleaching of the filter saturation of the gain starts to play a role, which, as is known [73], leads to shortening of the pulses. The effect of saturation is greater, the larger the gain of the active medium, i.e., the larger the initial density of the filter. Shortening of a pulse due to saturation continues until the pulse duration becomes comparable to T_2. From this moment on, effects of coherent interaction of radiation with matter start to play a role (in our case, the effect of coherent amplification), which also leads to shortening of the amplified pulse. Indeed, during coherent amplification formation of π-pulse occurs, for which we have the relation $W\tau_{pulse} = $ const [74, 75].

In our case the average energy W of the pulses for $\eta = 3\%$ and C = 50% was approximately 0.02 J, which for a pulse duration of 35 psec and beam cross section 3 mm^2 corresponds exactly to the energy of a π-pulse in ruby. It is reasonable to suppose that generation of a π-pulse occurred in our laser, and the shortening of the duration of the generated pulses as the initial density of the filter increased corresponds to temporal narrowing of the π-pulse during the process of its amplification.

In order to verify this hypothesis, we performed experiments on self-induced transparency and coherent amplification in ruby at a temperature of around 100°K using the output emission of our laser under conditions of self-mode-locking. The results of these experiments, discussed in the more recent papers [76, 77], indicate that generation of a π-pulse did indeed occur in our laser. Coherent effects in ruby at liquid nitrogen temperature are not considered here, since these questions lie outside the subject matter of this article.

In conclusion, the authors express their deep gratitude to M. D. Galanin for his interest in, and attention to, this work, as well as to E. D. Baivaya, O. P. Varnavskii, B. P. Kirsanov, Yu. N. Serdyuchenko, V. N. Smorchkov, Yu. S. Ushakov, and M. Ya. Shchelev for their cooperation and help with this work.

LITERATURE CITED

1. M. C. Adamson, T. P. Hughes, and K. M. Young, Electron. Quant., III, 2:1459 (1964).
2. F. A. Brand, H. Jacobs, and C. Locascio, Proc. IEEE, 52:1255 (1964).
3. S. Koozekanini, M. Gigtan, and A. Krutchkoff, Appl. Opt., 1:372 (1962).
4. K. Gurs, Phys. Lett., 16:125 (1965).
5. K. Gurs, Z. Naturforsch., 18a:418 (1963).
6. K. Gurs, J. Appl. Math. Phys., 16:49 (1965).
7. V. V. Korobkin and A. M. Leontovich, Zh. Éksp. Teor. Fiz., 49:10 (1965).
8. M. V. Danileiko, E. A. Tikhonov, and M. T. Shpak, Ukr. Fiz. Zh., 13:No. 3 (1968).
9. I. J. De'Haenens and C. K. Asawa, Electron. Quant., III, 2:1131 (1964).
10. I. J. De'Haenens and C. K. Asawa, Electron. Quant., III, 1:795 (1964).
11. V. K. Konyukhov, L. A. Kulevskii, and A. M. Prokhorov, Dokl. Akad. Nauk SSSR, 149:571 (1963).
12. A. Szabo, J. Appl. Phys., 35:2263 (1964).
13. V. K. Klinkov, Doctoral Dissertation, Kurnakov Institute of General and Inorganic Chemistry, Moscow (1972).
14. A. L. Schawlow, in: Advances in Quantum Electronics, J. R. Singer, ed., New York–London (1961), p. 50.
15. D. F. Nelson and M. D. Struge, Phys. Rev., 137:1117 (1965).
16. H. Statz and G. de Mars, in: Quantum Electronics, C. H. Townes, ed., Columbia Press, New York (1960), p. 150.
17. D. Ross, Proc. IEEE, 52:853 (1964).
18. R. J. Collins, D. F. Nelson, A. L. Schawlow, et al., Phys. Rev. Lett., 5:303 (1960).
19. J. P. Wittke, J. Appl. Phys., 33:2333 (1962).
20. A. M. Leontovich and V. L. Churkin, Zh. Éksp. Teor. Fiz., 59:7 (1970).
21. N. G. Basov, V. N. Morozov, and A. N. Oraevskii, Zh. Éksp. Teor. Fiz., 49:895 (1965); E. M. Belenov, V. N. Morozov, and A. N. Oraevskii, Trudy FIAN, 52:237 (1970); L. A. Ostrovskii, Zh. Éksp. Teor. Fiz., 48:1087 (1965); 49:1535 (1965).
22. A. F. Suchkov, Zh. Éksp. Teor. Fiz., 45:1495 (1965).
23. V. I. Bespalov and E. I. Yakubovich, Izv. Vyssh. Uchebn. Zaved., Radiofizika, 8:909 (1965).
24. G. L. Gurevich, L.Kh. Ingel', and Ya. I. Khanin, Kvantov. Élektron., No. 3(9):45 (1972).
25. G. I. Vonikurov, N. M. Galaktionov, V. F. Egorova, et al., Zh. Éksp. Teor. Fiz., 59:7 (1970).
26. B. P. Kursanov, A. M. Leontovich, and A. M. Mozharovskii, Kvantov. Élektron., 1:2211 (1974).
27. K. G. Folin, Doctoral Dissertation, Institute of Semiconductor Physics, Novosibirsk (1969).
28. A. T. Tursunov, Doctoral Dissertation, Kurnakov Institute of General Inorganic Chemistry, Moscow (1970).

29. A. M. Leontovich and M. N. Popova, Zh. Prikl. Spektrosk., 6:735 (1967).
30. V. V. Lyubimov and I. B. Orlova, Zh. Tekh. Fiz., 39:2183 (1970).
31. A. F. Suchkov, Trudy FIAN, 43:161 (1968).
32. R. I. Gintoft and A. M. Sarzhevskii, Dokl. Akad. Nauk Bel. SSR, 9:573 (1965).
33. V. L. Broude, V. I. Kravchenko, P. P. Pogerestkii, et al., Dokl. Akad. Nauk SSSR, 173:64 (1967).
34. A. S. Markin, Trudy FIAN, 56:3 (1971).
35. B. Ya. Zel'dovich and T. I. Kuznetsova, Usp. Fiz. Nauk, 106:47 (1972).
36. P. G. Kryukow and V. S. Letokhow, IEEE J. Quant. Electron., QE-8:766 (1972).
37. W. G. Wagner and B. A. Lengyel, J. Appl. Phys., 34:2040 (1963).
38. A. A. Grutter and H. Weber, Opto-electron., 3:13 (1971).
39. L. W. Riley, M. Bass, and E. L. Hahn, App. Phys. Lett., 7:88 (1965).
40. R. W. Williamson and D. Walsh, Proc. IEEE, 54:1122 (1966).
41. A. Szabo and L. E. Erickson, IEEE J. Quant. Electron., QE-4:692 (1968).
42. B. H. Soffer, J. Appl. Phys., 35:2551 (1964).
43. P. Kafalas, J. I. Masters, and E. M. Murray, J. Appl. Phys., 35:2349 (1964).
44. A. Szabo and R. A. Stein, J. Appl. Phys., 36:1562 (1965).
45. A. A. Kovalev, V. A. Pilipovich, A. A. Bogdanovskaya, et al., Zh. Prikl. Spektrosk., 9:71 (1968).
46. R. McLeary and P. W. Bowe, Appl. Phys. Lett., 8:116 (1966).
47. V. V. Korobkin, A. M. Leontovich, M. N. Popova, and M. Ya. Shchelev, Zh. Éksp. Teor. Fiz., 53:16 (1967).
48. V. S. Letokhov and A. F. Suchkov, Zh. Éksp. Teor. Fiz., 50:1148 (1966).
49. T. I. Kuzhetsova, Izv. Vyssh. Uchebn. Zaved., Radiofizika, 16:521 (1973).
50. V. D. Ivanov and A. M. Leontovich, Kvant. Élektron., No. 1:96 (1971).
51. H. W. Mocker and R. J. Collins, Appl. Phys. Lett., 7:270 (1965).
52. A. J. DeMaria, D. A. Stetser, and H. Heynay, Appl. Phys. Lett., 8:174 (1966).
53. V. I. Malishev and A. S. Markin, Zh. Éksp. Teor. Fiz., 50:339 (1966).
54. L. E. Hargrove and R. L. Fork, Appl. Phys. Lett., 5:4 (1964).
55. W. R. Sooy, Appl. Phys. Lett., 7:36 (1965).
56. S. D. Zakharov, P. G. Kryukov, Yu. A. Matveets, et al., Kvant Élektron., No. 5 (17):52 (1973).
57. P. G. Kryukov, Yu. A. Matveets, et al., Zh. Éksp. Teor. Fiz., 62:2036 (1972).
58. D. Von der Linde and K. F. Rodgers, IEEE J. Quant. Electron., QE-9:960 (1973).
59. M. A. Duguay and J. W. Hansen, Opt. Commun., 1:252 (1969).
60. M. W. McGeoch, Opt. Commun., 7:116 (1973).
61. P. G. Kryukov, Yu. A. Matveets, and O. B. Shatberashvili, Kvant. Élektron., 1:450 (1974).
62. M. C. Richardson, IEEE J. Quant. Electron., QE-9:768 (1973).
63. A. N. Zherikhin, P. G. Kryukov, E. V. Kurganova, et al., Zh. Éksp. Teor. Fiz., 66:116 (1976).
64. V. A. Babenko, V. I. Malishev, and A. A. Sychev, Pis'ma Zh. Éksp. Teor. Fiz., 14:461 (1971).
65. G. L. McAllister, M. M. Mann, and L. G. DeShaser, IEEE J. Quant. Electron., QE-6:44 (1970).
66. M. Hercher, W. Chu, and D. L. Stoockman, IEEE J. Quant. Electron., QE-6:44 (1970).
67. A. V. Milinkevich, V. A. Savva, and A. M. Samson, Zh. Prikl. Spektrosk., 22:997 (1975).
68. V. M. Baev, A. N. Savchenko, and E. A. Sviridenkov, FIAN Preprint No. 52 (1973).
69. A. P. Veduta, N. B. Fedotov, and N. P. Furzikov, Kvant. Élektron., 1:408 (1974).
70. V. A. Babenko, V. I. Malishev, A. A. Sychev, and A. N. Shibanov, Kvant. Élektron., 2:1923 (1975).

71.   G. Mourour, B. Drouin, M. Bergeron, and M. M. Denariez-Roberge, IEEE J. Quant.
      Electron., QE-9:745 (1973).
72.   M. L. Spaeth and W. R. Sooy, J. Chem. Phys., 48:2315 (1968).
73.   P. G. Kryukov and V. S. Letokhov, Usp. Fiz. Nauk, 99:169 (1969).
74.   S. L. McCall and E. L. Hahn, Phys. Rev., A2:861 (1970).
75.   S. L. McCall and E. L. Hahn, Phys. Rev., 183:457 (1969).
76.   A. M. Leontovich and A. M. Mozharovskii, Pis'ma Zh. Éksp. Teor. Fiz., 19:347 (1974).
77.   A. M. Leontovich and A. M. Mozharovskii, Pis'ma Zh. Éksp. Teor. Fiz., 20:664 (1974).

SPECTROSCOPY OF ACTIVATOR CENTERS OF RARE–EARTH IONS IN LASER CRYSTALS WITH GARNET STRUCTURE

Yu. K. Voron'ko and A. A. Sobol'

Spectroscopic characteristics of laser crystals with garnet structure containing rare-earth activators are analyzed. The mechanism of formation, properties, and statistics of different activator centers of the following ions are studied: isolated additive ions, combinations of additive ions, and proper structural defects. The effect of the structure and composition of the activator centers on the laser characteristics is studied for the materials investigated. Methods are discussed for analyzing structural imperfections in garnet crystals in terms of the spectra of rare-earth indicator ions introduced into the crystal. The material on the spectral properties of associates of rare-earth ions is used to study a mechanism for radiationless energy transfer effective only at the shortest distances between the rare-earth ions.

INTRODUCTION

Crystals with garnet structure activated by rare-earth (TR^{3+}) ions are at present widely employed as the active elements for solid lasers. Thus, the highest lasing powers for solid lasers in the cw regime at room temperature (to 300 W) have been obtained using crystals of aluminum—yttrium garnet with Nd^{3+}. The possibility of introducing into a garnet crystal various rare-earth ions (Nd^{3+}, Er^{3+}, Ho^{3+}, Tu^{3+}, Yb^{3+}) and obtaining stimulated emission from them makes it possible to develop garnet crystal lasers operating at different wavelengths [1-3].

Studies of the properties of special microvolumes of a crystal containing an ion-activator and its nearest-neighbor crystal enviroment (the so-called optical or activator centers [4-6]) have become widespread in the investigation of the causal relationship between the nature of change of the parameters of laser materials and the features of the structure of the host crystal and activator ions.

Since the so-called activator centers (AC) consist of interruptions of the periodicity of the crystal lattice with extremely small dimensions of the order of the lattice constant and less, spectroscopic methods are best suited for their study.

The application of these methods is due to the fact that properties such as the absorption and luminescence spectra, decay kinetics of excited states, and intensity of spectral lines of the optically active added ions are closely related with the symmetry and structure of the local neighborhood of these ions.

It should be noted that the spectroscopic studies of AC of rare-earth ions previously carried out were restricted mainly to fluorite crystals [4, 5, 7].

For garnet crystals not only is information on the formation of the AC of the rare-earth ions in this host crystal lacking, but it even remains unclear as to which structural defects may be present in such a matrix and how they might participate in the formation of various types of activator centers.

The main aim of this paper is to study the spectroscopy and character of formation of activator centers consisting of TR^{3+} ions in aluminum garnet crystals, and also to study the relation between the parameters of lasers based on these materials and the structure of the local crystal environments of the working ions.

In addition to studying the character of behavior of the additive TR^{3+} ions in crystals with garnet structure, the experiments performed to investigate the spectroscopy of associates of additive rare-earth ions made it possible to study a number of phenomena related to processes of energy transfer among these ions in an aluminum-garnet host crystal.

CHAPTER 1

LOCAL CRYSTAL ENVIRONMENTS OF TR^{3+} ION ADDITIVES IN CRYSTALS OF THE HOMOLOGOUS SERIES OF GARNETS $RE_3Al_5O_{12}$

§1. Symmetry of the Nearest-Neighbor Crystal Environment of an Isolated TR^{3+} Ion in a Garnet Lattice

The unit cell of compounds $RE_3Al_5O_{12}$ with garnet structure is cubic O_h^{12} [8, 9] (here RE^{3+} is the lanthanide series from Gd^{3+} to Lu^{3+}, including the ion Y^{3+}). We introduce the notation RE^{3+} for the rare-earth ions in order to distinguish the cases when these ions are components making up the garnet lattice from cases when the trivalent rare-earth cations are additives. For the latter case we will write TR^{3+}.

According to [9], the RE^{3+} ions occupy lattice points in the unit cell of garnet having a dodecahedral oxygen environment (d-sites), while Al^{3+} can appear at sites with octahedral (a) or tetrahedral (c) environments. The rare-earth ion additives occupy the site of an RE^{3+} cation. Hence, since the charges of the TR^{3+} additive and RE^{3+} cations are equal, the character of the electrostatic interaction between a cation at a d-site of the lattice with its environment does not change when an RE^{3+} cation is replaced by a TR^{3+} activator.

Nevertheless, there are factors which can lead to distortion of the original symmetry of a lattice d-site when an additive ion is present there, namely the difference in the ionic radii and the different structure of the electron shells of the RE^{3+} ion and the additive ion.

In this context it is of interest to ascertain whether such distortions are large, whether the special properties of individual additive ions affect the formation of their local environments, or whether the garnet lattice is so rigid that the role of such factor is virtually eliminated. These investigations are of interest in view of the fact that in a laser material such as $Y_3Al_5O_{12}-Nd^{3+}$, the sizes of the Y^{3+} and Nd^{3+} ions differ markedly.

The calculations carried out in [10-14] for the spectra of TR^{3+} additive ions using crystal field theory indicate that the best results are obtained when the crystal field potential corresponding to rhombic symmetry* is used in the calculations.

---

*There is clearly no need here to take into account the results of earlier work in which, in order to simplify the calculation, a higher symmetry type was taken for the crystal environment of a TR^{3+} ion in garnet [13].

We may thus conclude that when a cation of the host crystal is replaced by a TR^{3+} ion, the symmetry of the environment of a d-site is at least preserved. As for the effects of slight distortion of the structure, associated with some displacement of the nearest-neighbor ions of a TR^{3+} cation, the method of investigation based on calculation of the spectra [12-14] is too rough.

Therefore, in our investigations our aim was to ascertain the role played by specific features of the additive TR^{3+} ions in changing the local environment of an initial lattice site, and the following experimental technique was used. We did not study the absolute values of the Stark splitting of the energy levels of the TR^{3+} ions, but rather the way they changed in going from compound to compound in the homologous series of garnets $RE_3Al_5O_{12}$, where RE^{3+} was Tb, Dy, Ho, Er, Y, Tu, Yb, Lu. The results of x-ray analysis [9] indicate that all the compounds in this series have a cubic lattice with unit cells of the same structure. The crystal symmetry of the environment of a d-cation is also the same for all the garnets studied and corresponds to the orthorhombic group D_2. At the same time, in going from $Tb_3Al_5O_{12}$ to $Lu_3Al_5O_{12}$ in the garnet series the dimensions of the unit cell decrease, and therefore the interionic distances change. This in turn should cause a displacement of the position of the Stark components of the TR^{3+} ion when moving along the series of garnets. The way this type of displacements occurs will be determined by the corresponding changes in the interionic distances in the host crystal lattice under the condition that the structure of the matrix, and not the special features of the additive ion, principally determine the structure of the environment. In order to study the effect of the dimensions of the additive ion on the distortion of the structure of its local environment, Er^{3+}, Eu^{3+}, and Nd^{3+} activators were used, for which the ratios of the ionic radii (r) to the dimension of the cation RE^{3+} are different. Thus, the size of Eu^{3+} exceeds r for any RE^{3+} cation, the difference $r_{Eu^{3+}} - r_{RE}^{3+}$ increasing steadily as we move along the $RE_3Al_5O_{12}$ series. For Nd^{3+} this difference is even more significant. On the other hand, for Er^{3+}, when one RE^{3+} ion is replaced by another the value of $r_{Er^{3+}}$ can either exceed $r_{RE^{3+}}$ (for the interval from Er^{3+} to Lu^{3+}) or be smaller than $r_{RE^{3+}}$ (from Tb_3^{3+} to Ho^{3+}). That is, one expects the lattice distortions to be minimal for Er^{3+} when the latter is introduced into the crystal. The above additive TR^{3+} ions also have f-shells with differing structure. Thus as we go from Nd^{3+} to Er^{3+}, the f-shells become less extended due to the "lanthanide compression" effect [15]. It is possible to determine the role played by the size of the ion and the structure of its f-shell in the formation of the nearest-neighbor environment by studying the character of the changes of the positions of the components of the levels of these TR^{3+} ions in moving along the garnet series. The behavior of the displacement of the levels of the TR^{3+} ions, which should be determined by the change in the lattice structure of the crystal when one RE^{3+} cation is replaced by another, can be calculated theoretically using x-ray data. If the results of the calculations are then compared with the experimental curves, it is possible to answer not only the question of the extent to which alteration of the crystal environment of a TR^{3+} ion is determined by the parameters of the host crystal itself as we go along the $Re_3Al_5O_{12}$ series, but also to ascertain which changes of positions of the lattice ions are responsible for which aspects of the displacement of the components of the TR^{3+} ion levels. It should be noted that since we consider the way in which the position of the Stark components changes, and not the splittings of the energy levels, the application of crystal field methods, as will be shown below, does not involve the use of any quantities which decrease the accuracy of the calculations, in contrast to the situation in other papers.

In practice, it is convenient to work with the parameters of the crystal field B_n^l [16], the values of which can be obtained either by using the splitting of the energy levels or by making calculations from x-ray crystallographic data.

The procedure for calculating the parameters B_n^l from the experimentally observed level splitting is extremely laborious. However, for garnet crystals it turned out that only parameters of the form B_2^0 and B_2^2 need to be taken into account, since they are the ones which change the most as we move along the $Re_3Al_5O_{12}$ series. This conclusion was obtained by calculating the change in B^l in going from $Y_3Al_5O_{12}$ to $Lu_3Al_5O_{12}$ using the positions of the Stark sublevels $^4F_{3/2}$, $^4I_{\mathcal{I}}$ of the Nd^{3+} levels.

By [16], the Hamiltonian H for a rhombic symmetry has the form

$$H = \sum_{n=2}^{6} \sum_{l=0}^{n} B_n^l O_n^l, \tag{1}$$

where the O_n^l are the angular momentum operators [17]. A program was written for a BÉSM-4 computer which varied the values of the B_n^l in such a way as to obtain the best agreement between the positions of the Stark components determined experimentally and by theoretical calculation. The results of the calculation are shown in Table 1. It is seen from this table that as we go from $Y_3Al_5O_{12}$ to $Lu_3Al_5O_{12}$ it is indeed mainly the parameters with n = 2 which change. The values of these parameters can be easily obtained from the splitting of the energy levels of the TR^{3+} ions with total momentum values equal to 1 or $^3/_2$ using the following formulas which relate the positions of the levels E with B_2^l,

$$E_{1.2} = \pm \alpha [9 (B_2^0)^2 + 3 (B_2^2)^2]^{1/2} \quad \text{for } \mathcal{I} = {}^3/_2, \tag{2}$$

$$E_1 = -2\alpha B_2^0, \quad E_2 = \alpha (B_2^0 + B_2^2), \quad E_3 = \alpha (B_2^0 - B_2^c) \quad \text{for } \mathcal{I} = 1. \tag{3}$$

Here α is the equivalent operator constant [17]. In studying Nd^{3+} and Er^{3+} Eq. (2) must be used, while for Eu^{3+} Eq. (3) is valid. In order to be able to compare the experimental results obtained from the spectra of Er^{3+}, Nd^{3+}, and Eu^{3+}, i.e., to use Eqs. (2) and (3) simultaneously, it is convenient to define the function

$$p(z) = [3 (B_2^0)^2 + (B_2^c)^2]^{1/2},$$

where z is the atomic number of the RE^{3+} cation in the Lanthanide series. We then obtain for Er^{3+} and Nd^{3+} that

$$p(z) = \frac{\Delta E}{2\alpha \sqrt{3}},$$

while for Eu^{3+}

$$p(z) = \frac{1}{\alpha} \left[\frac{(\Delta E_1)^2 + (\Delta E_2)^2 - \Delta E_1 \Delta E_2}{3} \right]^{1/2}. \tag{4}$$

Here ΔE is the splitting of the doublet with $\mathcal{I} = {}^3/_2$; ΔE_1, ΔE_2 are the differences in the energies of the ground and excited sublevels of the triplet with $\mathcal{I} = 1$.

The values of α for Er^{3+}, Nd^{3+}, and Eu^{3+} were taken from [10, 12, 18].

Figure 1 shows the function p(z) for different states of Nd^{3+}, Er^{3+}, and Eu^{3+}. It is seen from the figure that all the curves except the one for p(z) for the level $^4S_{3/2}$ of Nd^{3+} change in the same way, i.e., they do not depend on the values of the orbital momentum (L) of the states considered or on the position of the TR^{3+} additive in the lanthanide series. Hence no difference

TABLE 1

B_n^l	B_2^0	B_2^2	B_4^0	B_4^2	B_4^4	B_6^0	B_6^2	B_6^4	B_6^6
$Y_3Al_5O_{12}$	±150	±230	-100	-186	-1100	30	—	-635	802
$Lu_3Al_5O_{12}$	±100	±187	-91	-180	-1090	30	—	-635	810

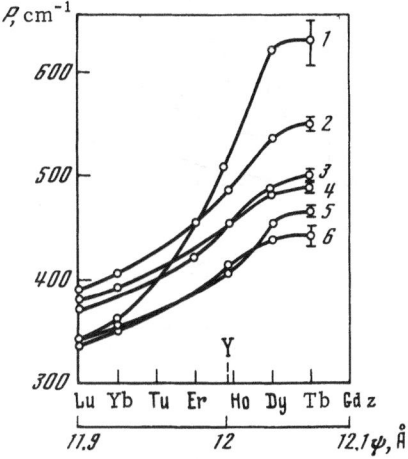

Fig. 1. Dependence of p(z) on the atomic number of the RE^{3+} cation and the lattice constant (ψ) for members of the $Re_3Al_5O_{12}$ garnet series for the levels 1) $^4S_{3/2}Nd^{3+}$; 2) $^4F_{3/2}Nd^{3+}$: 3) $^4S_{3/2}Er^{3+}$; 4) $^7F_1Er^{3+}$; 5) $^5D_1Eu^{3+}$; 6) $^4F_{3/2}Eu^{3+}$.

in the change of p(z) is observed for Nd^{3+}, Eu^{3+}, and Er^{3+}, which have substantially different sizes and f-shell structures. It can be concluded here that the change of p(z) is mainly caused by change of the crystallographic parameters of the host crystal, and the individual properties of the TR^{3+} ions play practically no role. The peculiar behavior of the curve p(z) for the $^4S_{3/2}$ level in Nd^{3+} can be explained easily if the following fact is taken into account. States with L = 0 (S-states) can be split by the crystal field only when other states with L ≠ 0 are mixed with the S-state as a result of interbond coupling (spin-orbit mixing). The amount of spin-orbit mixing should change if the f-shell of the TR^{3+} ion is subjected to a symmetric compression. This effect is possible for Nd^{3+}, the radius of which is larger than that of any RE^{3+} cation, and as we go along the series of garnets, the amount of the compression will change as the ratio r_{RE}/r_{Nd} changes.

Consequently, along with a change of the parameters of the host crystal, the degree of spin-orbit mixing of the states of Nd^{3+} will also vary. As already noted, this effect is strongest for the $^4S_{3/2}$ level and does not affect the $^4F_{3/2}$ level, since the splitting of this level (here L ≠ 0) is not determined by the effect of mixing of states. Since the ionic radius of Er^{3+} corresponds to the dimensions of an RE^{3+} cation, its f-shell will not undergo compression and hence there is no change in the spin-orbit mixing as we go along the $Re_3Al_5O_{12}$ series. For this reason the behavior of p(z) for the level $^4S_{3/2}$ of Er^{3+}, in contrast to Nd^{3+}, behaves like the other p(z) curves in Fig. 1.

The function p(z) which we use is given by some combination of the parameters B_2^0 and B_2^2 [Eq. (4)]. It is convenient to study the change of these parameters over the series $Re_3Al_5O_{12}$ using the example of the splitting of states with $\mathcal{F} = 1$ [Eq. (3)]. Since the symbols E_1, E_2, and

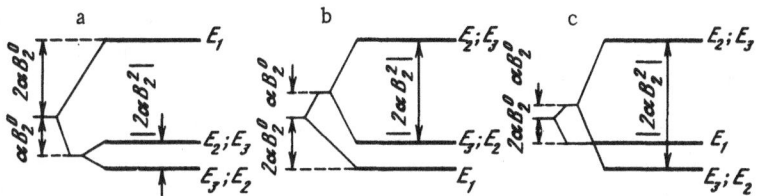

Fig. 2. Possible splitting patterns of levels with $\mathcal{F} = 1$ for B_2^0 of different sign and different ratio B_2^0/B_2^2.

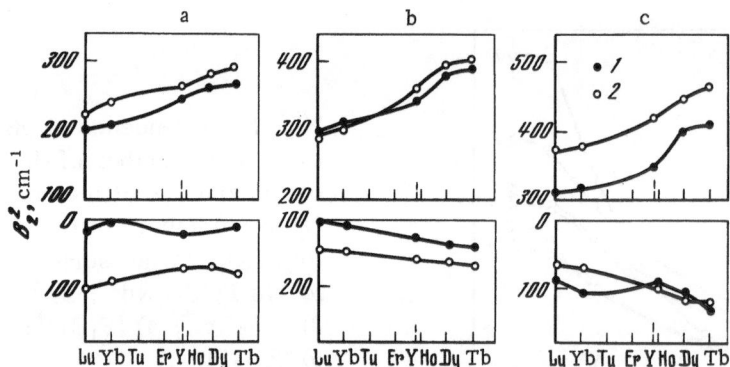

Fig. 3. Dependence of B_2^0 and B_2^2 on the atomic number of
the RE^{3+} cation for three splitting patterns of the level
with $\mathcal{J} = 1$ (a, b, and c correspond to Fig. 2). a) B_2^0 top,
B_2^2 bottom; b) B_2^2 top, B_2^0 bottom, 1) $^5D_1$, 2) $^7F_1$.

E_3 in (3) may be assigned to the three Stark components of the level $\mathcal{J} = 1$ in three different
ways (Fig. 2), three sets of parameters B_2^0 and B_2^2 can be obtained experimentally. It should be
noted that for the case of D_2 symmetry, all three sets of parameters B_2^2 will be regular and each
choice corresponds to a definite choice of quantization axis along one of the three two-fold axes.
All three choices transform into one another when the quantization axis is changed. It is possible to verify here whether in fact there exists a TR^{3+} ion environment having the exact symmetry D_2 or not by constructing the functions $B_2^0(z)$ and $B_2^2(z)$ for the three possible choices of
these parameters and using experimental data on the splitting of levels with the same $\mathcal{J}$ but
different L ($^5D_1$, $^7F_1 Eu^{3+}$). These functions are shown in Fig. 3. It is seen from the figure that
the behavior of B_2^0 and B_2^2 for the levels $^5D_1$, $^7F_1$ coincides for only one choice of the parameters
(case b); if the symmetry of the environment were exactly D_2, the same behavior would be observed for all three choices. It is also seen from Fig. 3 that the most unfavorable case occurs
for $B_2^0 > 0$ and $B_2^2 = \min$ (case a), which in Koningstein's paper [10] was taken as the basis for the
investigation of the crystal field of TR^{3+} ions in garnet crystals. In his later investigations by
the method of electron Raman scattering [19], Koningstein disavowed his initial interpretation
and took the case $B_2^0 < 0$ to be the basis of the investigation, i.e., he also arrived at a result
coinciding with our investigations.

Thus it follows from Fig. 3 that there exists only one choice of the parameters of the
Hamiltonian in (1) which best describes the experimental results and not three choices, which
would be the case if there were exact D_2 symmetry. Hence the symmetry of the local environment of a TR^{3+} ion does not coincide exactly with the symmetry obtained from the results of
x-ray analysis. This discrepancy can clearly not be ascribed to deformation of the environment
as a result of introducing a TR^{3+} ion into the d-sublattice, since in this case there would be a
dependence of the results on the size of the additive ion, which in actuality is not observed. The
most likely explanation is that the environment of the TR^{3+} ion does not correspond exactly to
the model corresponding to the data from x-ray analysis.

However, it should be noted that there is no gross contradiction with the x-ray data concerning the symmetry of a d-site in a garnet lattice, since the observed divergence of the results reduces only to requiring a definite choice of the quantization axis of the Hamiltonian of
form (1), and the number of parameters B_n^l must correspond to rhombic symmetry, as follows
from the results of x-ray analysis.

In this context the use of results from x-ray analysis is not ruled out as a first approximation for calculating the parameters B_n^l. These calculations were performed by us with the aim of explaining the peculiar properties of the behavior of $B_2^0(z)$ and $B_2^2(z)$.

As is seen from Figs. 1 and 3, the curves $B_2^0(z)$ and $B_2^2(z)$ behave in the following way: They increase with increasing lattice constant, and they have a discernible kink in the region $Y_3Al_5O_{12} - Tb_3Al_5O_{12}$.

It is known from [20] that

$$B_n^l = A_n^l \langle r^n \rangle,$$
$$A_n^l = \frac{2\pi}{2n+1} \sum_i \frac{eq_i}{R_i^{n+1}} Y_n^{l*}(Q_i, \varphi_i).$$

(5)

Here $\langle r^n \rangle$ is a radial integral; R_i, Q_i, φ_i are the spherical coordinates of the i-th ion in the lattice (Fig. 4); q_i is the charge of the i-th ion. The value of q_i was assumed equal to the normal valence. The lattice sums were performed over spheres of differing radii out to R = 12 Å. It was most convenient to calculate the curve p(z), since it is easy to show that no difficulties arise with the choice of the quantization axis in this case. In addition, in order to eliminate the parameter $\langle r^n \rangle$, which is not accurately known, the curve $\beta = p(z)/p(Y)$ was calculated.

At first glance, Eq. (5) implies that as we go along the series in the direction $Lu^{3+} \rightarrow Tb^{3+}$ we should get an R^{-3} dependence for B_2^0, B_2^2, p(z), although as seen from Figs. 1 and 3, the opposite dependence is observed.

Calculation of the functions B_2^0, B_2^2, p(z) using Eq. (5) reveals, however, that the peculiar behavior of the experimental curve for p(z) does not contradict the change of the crystallographic parameters of the host crystal as we go along the $Re_3Al_5O_{12}$ series. It is seen from Fig. 5 that there is good agreement of the experimental and calculated curves for p(z) at R = 3.7 Å (i.e., the first coordination sphere). Allowance for the contribution to the lattice sum from the ligand charge located at a greater distance leads to a dependence of p(z) differing markedly from the experimental dependence. It can therefore be assumed that the dependence of p(z) is determined only by the contribution to the sum (5) of the first coordination sphere of the TR^{3+} ion environment. This assumption is indeed possible if it is taken into account that the values of the effective charges q_i may decrease significantly due to screening of the distant ions by the innermost oxygen environment. In addition, the contribution from the Al^{3+} ions may also be significantly reduced owing to the partially covalent Al−O bond.

On the basis of this assumption, the peculiar behavior of p(z) can be explained by the features of the garnet crystal structure, as follows.

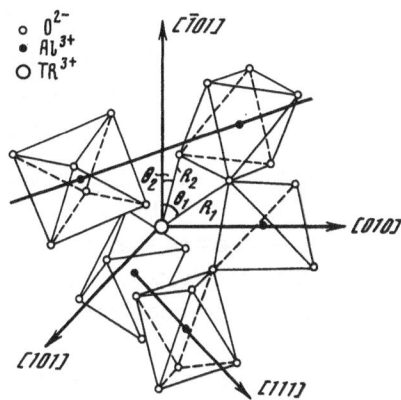

Fig. 4. Structure of the nearest-neighbor environment of a TR^{3+} ion in a garnet lattice.

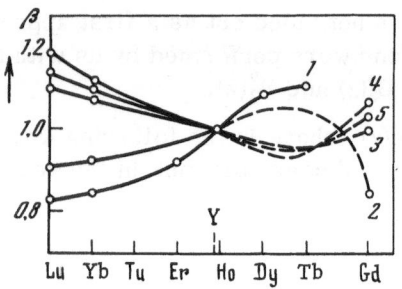

Fig. 5. Dependence of $\beta(z) = p(z)/p(Y)$ on the atomic number of the RE^{3+} cation. 1) Experimental curve; 2) curve calculated by Eq. (5) with summation over the first coordination sphere; 3) with summation over a sphere of radius $R = (\psi/4)\sqrt{3}/2$; 4) with summation over the sphere $R = \psi/2$; 5) with summation over the sphere $R = \psi$.

When we go from $Lu_3Al_5O_{12}$ to $Y_3Al_5O_{12}$, the quantities R_1, R_2 (cf. Fig. 5) increase in accordance with the increase of the lattice constant, which should decrease B_2^0, B_2^2 if we ignore the change in the angular parts Y_n^l of the parameters [Eq. (5)]. However, along with the increase in R, the quantities Y_n^l in Eq. (5) increase simultaneously due to an increase of the angles Q_i, φ_n, which is associated with compression of the Al—O octahedra (cf. Fig. 5) along the third-order axis. The last effect more than compensates for the reduction of B_2^0, B_2^2 due to the increase of R_i and causes them both to increase.

In the region from $Y_3Al_5O_{12}$ to $Gd_3Al_5O_{12}$, the growth of Y_n^l is retarded because there is no compression of the Al—O octahedra noted above in this region, and hence the uncompensated growth of the quantities R_i leads to a kink of the curve $p(z)$ here.

Thus the above investigations reveal that in a rigid lattice of a garnet of type $RE_3Al_5O_{12}$ the individual properties of the TR^{3+} additive ions do not manifest themselves. The structure of the local environment of the TR^{3+} additive and the way in which it changes are determined principally by the structure of the crystal matrix, in which a special role is played by the structure of the Al—O complexes.

§2. Classification of Activator Centers of TR^{3+} Ions

in $Y_3Al_5O_{12}$ Crystals

As follows from the preceding section, the TR^{3+} ions are located at d-sites of the lattice, and in the first approximation their nearest-neighbor structure is the same as for the RE^{3+} cations, provided that additive ions of other species are not present in neighboring sites. This case corresponds to a situation in which there exists only one type of AC in the crystal.

Additional additives have recently been introduced into $Y_3Al_5O_{12}$ crystals along with the working ions Nd^{3+}. One of the reasons for developing such materials, which are called mixed garnets, was the need to increase the concentration of Nd^{3+} and its distribution coefficient, which, due to the large difference in the ionic radii of Nd^{3+} and Y^{3+} in $Y_3Al_5O_{12}$, does not exceed 0.2.

In order to increase the distribution coefficient of Nd^{3+}, a large number of Gd^{3+} ions (up to 20 at. %) was introduced into crystals along with Nd^{3+}. As a result of the increase in the dimensions of the unit cell, the distribution coefficient of Nd^{3+} increased by 30%.

Another way to increase the distribution coefficient is to choose a coactivator which would compensate for the difference in the sizes of the main additive and the cation in the host crystal [22]. In [22] it was suggested that Lu^{3+}, which in contrast to Nd^{3+} has dimensions smaller than Y^{3+}, be used as the coactivator. According to the data of [22], in crystals with "compensated volume" the distribution coefficient of Nd^{3+} increased by 50%. Subsequent investigations showed that this claim was, however, in error [23]. Nevertheless, the opinion exists that the introduction of Lu^{3+} together with Nd^{3+} improves the characteristics of laser material.

It is easy to see that the presence of additional additives in mixed garnet crystals extends the possibilities for forming different types of AC of the working ions, due to the appearance of associates involving the principal and the secondary additives.

It has been suggested that the effect of broadening of the spectral lines of Nd^{3+} in the mixed crystals $(YLu)_3Al_5O_{12}$ [24] and $Y_3(Al, Ga)_5O_{12}$ [25, 26] be used to increase the efficiency of laser materials operating under Q-switching conditions.

We pursued two aims in our investigation of mixed garnet crystals. The first aim was to study the mutual effect of rare-earth ions with different sizes on the character of formation of associates among these ions. The second aim was to ascertain the most efficient ways to influence the width and intensity of the spectral lines of the working ions. These problems could only be solved by studying the general mechanism for the formation of activator centers in mixed garnet crystals. The investigations were based on studying the spectra of indicator TR^{3+} ions specially introduced into the crystal. For convenience, we call the TR^{3+} ions whose spectra are used to record the changes in the nearest-neighbor environment A^{3+} ions. These indicator ions were introduced in small concentration into the crystal in order to prevent formation of associates of A^{3+} ions. In contrast to the A^{3+} ions, the TR^{3+} ions which were added to the crystals as secondary additives (the case of mixed rare-earth garnets) will be denoted as RE^{3+} ions.

We now consider possible models of activator centers, starting from the crystallographic structure of garnet.

1. The case of mixed rare-earth garnets $(Y_{1-x}RE_x)_3Al_5O_{12}$($RE^{3+}$ is any rare-earth ion in the series from Gd^{3+} to Lu^{3+}). In this case we have an isovalent replacement of Y^{3+}. The A^{3+} and RE^{3+} ions are located at the d-sites of the lattice. The nearest-neighbor structure of A^{3+} for this case is shown in Fig. 6. On the basis of this figure, the general formula for the possible compositions can be written as

$$A^{3+} + nRE^{3+} \quad (n = 0 \text{ to } 4). \tag{6}$$

It is easily calculated that the total number of AC types for this case is equal to seven. The compositions with n = 0, 4, 1, and 3 correspond to one type of AC each, and n = 2 corresponds to three inequivalent AC.

The analysis of the AC in rare-earth garnet crystals used $(Y_{1-x}RE_x)_3Al_5O_{12}$ samples with x = 0-1, RE = Er, Ho, Tu, Yb, Lu, in which the role of A^{3+} ions was played by Er^{3+}, Nd^{3+}, Pr^{3+}, Eu^{3+}.

2. Mixed garnets of composition $Y_3Al_5O_{12} - L_3Al_2R_3O_{12}$ (L^{2+} divalent, R^{4+} tetravalent cations). This case differs from the one considered above in the heterovalency of the substituted additives, and also in that replacement of cations occurs both in the yttrium and aluminum sub-

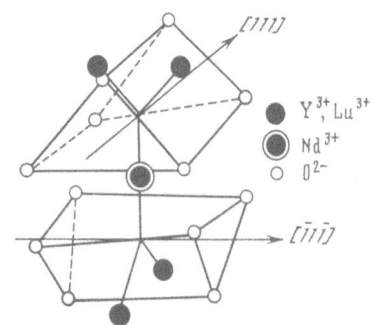

Fig. 6. Model of the local environment of Nd^{3+} in a mixed $(Y_xLu_{1-x})_3Al_5O_{12}$ garnet.

lattices (c-sites). A general formula for the composition of an activator site in this type of compound can be written as

$$A^{3+} + nL^{3+} + pR^{4+} \quad (n = 0 \text{ to } 4; \quad p = 0 \text{ to } 2); \tag{7}$$

the total number of AC in this type of mixed garnets is significantly greater than the number of AC for the case considered above. The crystals of the class of compounds considered had the composition $Y_3Al_5O_{12} - L_3Al_2Si_3O_{12}$ ($L^{2+} = Mg^{2+}$, Mn^{2+}, Ca^{2+}) with a concentration as large as 10 mol.%. Due to the large sizes of Mn^{2+}, Ca^{2+} (1.01; 0.91 Å)* their most likely location is at the d-sites of the lattice, whereas Mg^{2+} (ionic radius 0.74 Å) can replace either Y^{3+} or Al^{3+}. Tetravalent Si^{IV} compensating the charge of the divalent cations must replace Al^{3+} at the tetrahedral sites, since it is improbable for Si^{IV} to have a coordination number greater than four [28].

3. The case of isovalent replacement of Al^{3+} in $Y_3Al_5O_{12}$ by R^{3+} cations. Figure 4 gives an idea of the model for this type of AC. On the basis of Fig. 4, and also taking into account that Al^{3+} has two inequivalent positions, the formula for the composition of an activator center can be written as

$$A^{3+} + pR^{3+} + qR^{3+} \quad (p = 0 \text{ to } 2; \quad q = 0 \text{ to } 4). \tag{8}$$

Sc^{3+} ions (in concentrations up to 10 at.%) were used as the additive R^{3+} ions in the crystals studied. That the Sc^{3+} ions are preferentially located at the aluminum sublattice sites can be confirmed by the results of x-ray analysis. As was shown by our investigations, the size of the unit cell (ψ is 12 Å) increases to 12.012 Å in $Y_3Al_5O_{12}$ to which 8 at.% Sc^{3+} has been added. This result is possible if Al^{3+} is replaced by Sc^{3+}, which is larger. If Sc^{3+} replaced Y^{3+}, the dimensions of the unit cell would decrease since the ionic radius of Sc^{3+} (0.82 Å) is less than that of Y^{3+} (0.91 Å). As shown in [5], the amount of various types of AC is determined by the concentration of additive ions and by the way in which they are distributed over the lattice sites. An equiprobable distribution can be expected for cases (6) and (8) since they involve isovalent replacement of the additive and the role of electrostatic forces in the formation of associates of composition (1) and (3) is small. However, this supposition requires confirmation, since a departure from an equiprobable distribution is possible due to the appearance of elastic or covalent forces. In particular, it is possible for the case of "compensated volume" to be realized, as pointed out in [22]. As for case (2), electrostatic attraction of the L^{2+} and R^{4+} cations is possible here. Such an attraction can facilitate the formation of AC (composed in accordance with Eq. (7) of both L^{2+} and R^{4+} ions simultaneously) when the concentrations of the additives in the crystal are much less than when they are equiprobably distributed.

The character of the distribution of the additive ions over the lattice sites can be determined by comparing the experimental dependences of the number of centers of a given type on the total concentration of additive with the dependences determined by calculation under the assumption of equiprobable distribution. The calculated curves can be constructed using the equation

$$C_n = \frac{k!}{n!\,(k-n)!}\,\beta_1\beta_2^n\,[1 - (\beta_1 + \beta_2)]^{(k-n)} \tag{9}$$

from probability theory [29]. Here C_n is the concentration of A^{3+} cations having n neighboring ions of another additive in the innermost coordination sphere, and n must be $\leq$ k; k is the co-

---

*The preferential distribution of Mn^{2+} at the d-sites was reported in [27] on the basis of results of EPR studies of garnets of the above type.

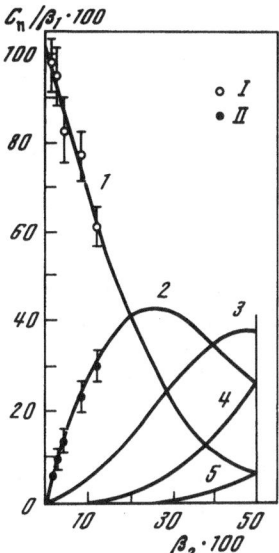

Fig. 7. Calculated dependences of
the relative concentration of A^{3+}
ions having n neighboring RE^{3+} ions
in the first coordination sphere on
the total concentration of RE^{3+} ions
in the crystal for the following cases:
1) n = 0; 2) n = 1; 3) n = 2; 4) n = 3;
5) n = 4; I and II, experimental dependences of the change of concentration
of N- and M-centers.

ordination number (in our case, 2 or 4); β is the relative concentration of A^{3+} ions; β_2 is the
relative concentration of the ions of the supplementary additive.

Figure 7 shows graphs of C_n/β_1 for the case k = 4.

§3. Analysis and Properties of AC of TR^{3+} Ions
in Mixed Garnets of Different Composition

a. Crystals of mixed rare-earth garnets $(Y_{1-x}RE_x)_3Al_5O_{12}$. A comparison of the absorption and luminescence spectra of A^{3+} ions in the mixed garnets $(Y_{1-x}RE_x)_3Al_5O_{12}$ and in
the garnet $Y_3Al_5O_{12}$ reveals that they are identical for the most part. Disordering of the structure of the compounds $(Y_{1-x}Lu_x)_3Al_5O_{12}$ causes only a slight broadening (by $\approx$ 2-10 times at
77°K) and displacement of the spectral lines ($\approx$ 1-5 cm^{-1}) of the A^{3+} ions compared to crystals
of $Y_3Al_5O_{12}$. These facts are illustrated in Fig. 8 for Nd^{3+} in crystals of the series of solid
solutions $(Y_{1-x}Lu_x)_3Al_5O_{12}$ in which x varies continuously (for the $^4I_{9/2} \rightarrow {}^4F_{3/2}$ transition at
77°K). In mixed garnet crystals $(Y_{1-x}Lu_x)_3Al_5O_{12}$ there was also no change of the intensities

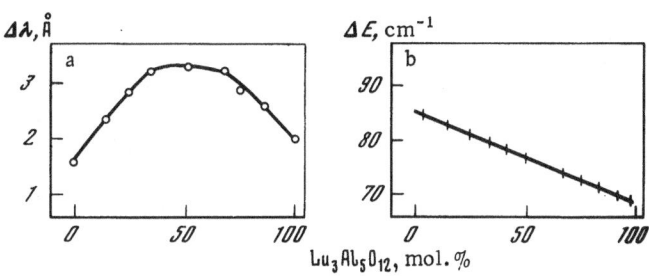

Fig. 8. Dependences of the width of the absorption
line corresponding to a transition between the
lower Stark components of the levels $^4I_{9/2} \rightarrow {}^4F_{3/2}$
of Nd^{3+} (a) and the value of the splitting of the
level $^4F_{3/2}$ on the value of x in $(Y_{1-x}Lu_x)_3Al_5O_{12}$
at 77°K (b).

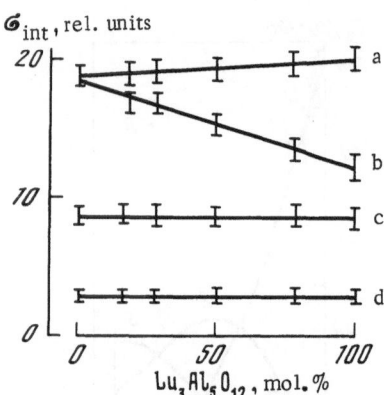

Fig. 9. Dependence of the integrated line absorption section in the group $^4I_{9/2} \rightarrow {}^4F_{7/2}$ (a, b) and in the group $^4I_{9/2} \rightarrow {}^4F_{3/2}$ (c, d) on the value of x in $(Y_{1-x}Lu_x)_3Al_5O_{12}$ at 77°K.

recorded in general of the spectral lines with increasing x. It was possible to find only a few transitions in Nd^{3+} where the value of the integrated absorption cross section varied as x changed in the series $(Y_{1-x}Lu_x)_3Al_5O_{12}$, and the range of variation did not exceed 30–40%. These cases are illustrated in Fig. 9. For the most part, the transition probabilities of the lines of the Nd^{3+} ions remained constant to within 10%.

An investigation of the character of the inhomogeneous broadening of the lines gives little information on the structure and properties of the activator centers in this class of crystals. It is therefore necessary to find transitions where it is possible to record the inhomogeneous line splitting, which would make it possible to distinguish the spectra belonging to particular types of AC. A thorough investigation revealed that the number of such transitions is small. Inhomogeneous splitting was observed at 77 and 4.2°K for the following transitions: $^5D_0 \rightarrow {}^7F_1$ in Eu^{3+} and $^4F_{3/2} \rightarrow {}^4I_{11/2}$ in Nd^{3+} in the luminescence and $^3H_4 \rightarrow {}^3P_1$ for Pr^{3+} in the absorption spectrum. It is apparent that all the above transitions are observed between levels, for one of which $\mathscr{I} = {}^3/_2$ or 1. This fact indicates that the second-order parameters of the crystal field potential are subject to the greatest variation in mixed garnet crystals, which is in accord with the conclusions of Chapter 1, Sec. 1.

Let us consider the features of the inhomogeneous splitting and broadening of the luminescence line in the group $^4F_{1/2} \rightarrow {}^4I_{11/2}$ Nd^{3+} at 77°K as x changes in the series $(Y_{1-x}Lu_x)_3Al_5O_{12}$ (Fig. 10). It follows from this figure that in the range of Lu^{3+} concentrations from 2 to 25 at.%,

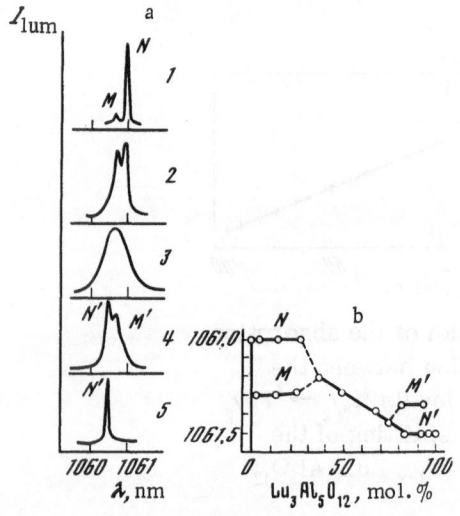

Fig. 10. Fine structure of the luminescence line corresponding to a transition between the lower levels of the Stark components of the levels $^4F_{3/2}$, $^4I_{11/2}$, (a) and the position of the principal and satellite lines (b) in crystals $(Y_{1-x}Lu_x)_3 \cdot Al_5O_{12}$ of different composition at 77°K. 1–5 correspond to the following concentrations of $Lu_3Al_5O_{12}$ in mol.%: 1) 3; 2) 13; 3) 50; 4) 82; 5) 97.

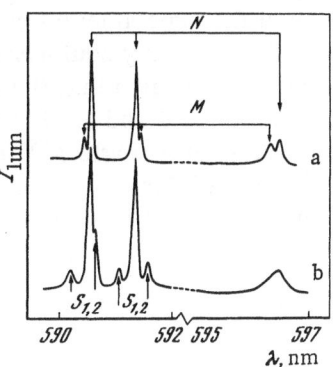

Fig. 11. Luminescence spectrum of Eu^{3+} for the transition $^5D_0 \rightarrow {}^7F_1$ in $Y_3Al_5O_{12}$-Lu^{3+} (10 at.%) (a) and in $Y_3Al_5O_{12}$-Sc^{3+} (8 at.%) (b) at 4.2°K.

there appears in the spectrum a satellite line M, in addition to the principal line which is usually observed in $Y_3Al_5O_{12}$ (the N-line). As the concentration of Lu^{3+} increases, so does the relative intensity of the M-line, and at 25 at.% Lu, both lines coalesce. In this case the width of the N-line is 8-9 times greater than the width of the line for Nd^{3+} in $Y_3Al_5O_{12}$. When the concentration of Lu^{3+} increases further, the center of the broadened line is linearly displaced toward shorter wavelengths, and at 82 at.% Lu, the line again splits, but now into N'- and M'-lines, the N'-line having the same position as the corresponding line for Nd^{3+} in $Lu_3Al_5O_{12}$. The same situation for the inhomogeneous splitting of the spectral lines was also observed in the spectra of Eu^{3+} and Pr^{3+} (Fig. 11a and 12b).

We can thus conclude that in the ranges of compositions corresponding to Lu^{3+} concentrations up to 25 and greater than 80%, two types of AC are mainly present in the crystals: N, M or N', M'. For concentrations in the range 26-80 at.% the fine structure of the A^{3+} lines could not be observed.

In order to determine to which type of AC the N- and M-lines belong, as well as to find the statistics of the centers, graphs were constructed of the relative concentrations of the A^{3+} ions appearing in each center, $\alpha = C_M/(C_N + C_N)$ and $\zeta = C_M/(C_N + C_N)$, as functions of the total Lu^{3+} concentration (β_2). The concentrations C_N and C_M were found from the values of the integrated absorption coefficients of the N and M lines.

These graphs are shown in Fig. 7 for the case of the absorption spectrum of Pr^{3+} in the group $^3H_4 \rightarrow {}^3P_1$ at 4.2°K. The same results were obtained when the A^{3+} ions studied were Nd^{3+} or Eu^{3+}. The concentrations of the A^{3+} ions and the Lu^{3+} ions in the crystal were determined by the method considered in [30].

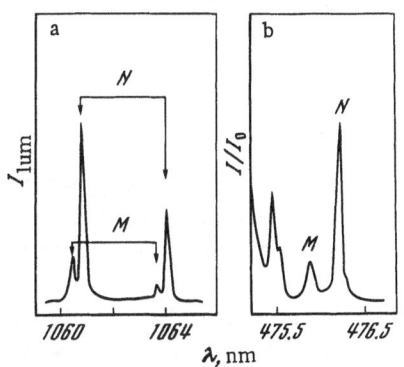

Fig. 12. Luminescence spectrum of Nd^{3+} in the group $^4F_{3/2} \rightarrow {}^4I_{11/2}$ at 4.2°K (a) and absorption spectrum of Pr^{3+} in the group $^3H_4 \rightarrow {}^3P_1$ at 77°K in crystals of $Y_3Al_5O_{12}$-Lu^{3+} (10 at.%).

A comparison of the experimental curves for $\alpha(\beta_2)$ and $\zeta(\beta_2)$ with the dependences C_n/β_1 given by Eq. (9), assuming equiprobable distribution of the A^{3+} and Lu^{3+} ions over the sites of the d-sublattice, revealed that the dependences of α and ζ agree well with the calculated curves C_n/β_1 for n = 0 and 1. These results indicate that the N- and M-centers can be modelled as consisting of a single A^{3+} ion and pair $A^{3+} + Lu^{3+}$, respectively, and the distribution of the A^{3+} and Lu^{3+} ions over the lattice sites is equiprobable. The model of an isolated A^{3+} ion for an N-center is also supported by the fact that in this type of AC the positions of the A^{3+} line coincide exactly with the positions of the A^{3+} lines in $Y_3Al_5O_{12}$.

It follows from an analysis of the graphs of C_n/β_1 shown in Fig. 7 that the situation in which there exist two primary types of activator centers in the crystal is preserved up to a Lu^{3+} concentration of 20-25 at.%. This explains the splitting of the lines into two components N and M observed by us in the spectra of the A^{3+} ions in $(Y_{1-x}RE_x)_3Al_5O_{12}$. When the concentration β_2 increases further, the number of centers corresponding to the models of centers given by Eq. (6) with n > 1 becomes large, as follows from Fig. 7, and it is not possible to distinguish the lines of the individual AC due to their large number. When the concentration of Lu^{3+} becomes larger than 80 at.%, a situation arises in the crystal which is similar to the case when $\beta_2 < 20$ at.%, with the difference that the isolated A^{3+} ion has an environment similar to that of the A^{3+} ion in $Lu_3Al_5O_{12}$, and the pair has the composition $A^{3+} + Y^{3+}$.

The fact demonstrated by us that ions Nd^{3+} and Lu^{3+}, for which by [22] a "compensation of volume" effect should be observed, have equiprobable location over the lattice sites, indicates that the effect of elastic interaction forces is extremely small on the position of the additive ions in the $Y_3Al_5O_{12}$ lattice. The reason for this phenomenon could be the significantly high degree of rigidity of the lattice which was shown in Sec. 1 to be characteristic of aluminum garnet crystals. As a result of this rigidity, the effective radius of elastic forces is limited.

Since an effect of one type of ion on the concentration of another type when they are both introduced together into a crystal can only occur when the two types of ion interact [31], the small effective radius of elastic forces in $Y_3Al_5O_{12}$ even at minimal distances between the cations in the lattice allows one to conclude that it is not possible to increase the distribution coefficient of certain rare-earth ions (Nd^{3+}) by introducing other ions (Lu^{3+}) into the crystal as was proposed in [22].

We now consider the relation between the size of the shift of the spectral M-line relative to the N-line and the specific properties of the RE^{3+} ions introduced into the crystal. The maximal shift $\Delta = 3-5$ cm^{-1} was observed for the case when $RE^{3+} = Lu^{3+}$, after which it decreases as the value of the ionic radius of the RE^{3+} ion approached that of the Yu^{3+} ion. Thus for the cases $RE^{3+} = Ho^{3+}$, the value of the ratio $r_{Ho^{3+}}/r_{Y^{3+}}$ is nearly one. In these crystals, the N- and M-lines practically coalesce. These facts indicate that in the absence of electrostatic interaction between the A^{3+} and RE^{3+} ions, the perturbation of the crystal environment of an isolated A^{3+} ion when another rare-earth ion is present at an adjacent d-site of the lattice is associated with the local deformation of the lattice. The size of this perturbation is determined by the ratio of the ionic radii of the Y^{3+} and RE^{3+} cations and cannot be significant in aluminum garnet crystals (this follows from the small displacement of the lines of the A^{3+} ions appearing in the associates $A^{3+} + RE^{3+}$).

b. In this connection mixed garnets of composition $Y_3Al_5O_{12} - L_3Si_3Al_2O_{12}$, where $L^{2+} = Mg^{2+}, Mn^{2+}, Ca^{2+}$, are of interest. In contrast to the case considered above, these garnets made it possible to study the special features of formation of centers for the case of heterovalent replacement of ions in the yttrium and aluminum sublattices. Figure 13a, b shows the situation for inhomogeneous splitting of the absorption spectra for Nd^{3+} in the group $^4I_{9/2} \rightarrow {}^4F_{3/2}$ at 4.2°K for this type of crystals. It follows from Fig. 13 that the spectra of the crystals for which $L^{2+} =$

Fig. 13. Absorption spectrum of Nd^{3+} (0.05 at.%) for the transition $^4I_{9/2} \rightarrow {}^4F_{3/2}$ at 4.2° K in the following crystals: a) $Y_3Al_5O_{12}-Mg_3Si_3Al_2O_{12}$ (8 mol.%); b) $Y_3Al_5O_{12}-Mn_3Si_3Al_2O_{12}$ (8 mol.%), $Y_3Al_5O_{12}-Ca_3Si_3Al_2O_{12}$ (8 mol.%); c) $Y_3Al_5O_{12}-Sc^{3+}$ (10 at.%).

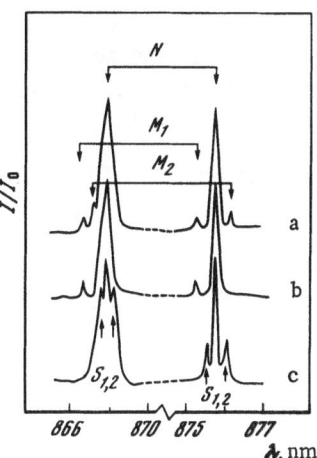

Mn^{2+}, Ca^{2+} are identical and are characterized by the presence of two types of AC: N, M_1; but in a crystal with Mg^{2+} additive, in addition to the N- and M_1-lines, M_2-lines are present. The displacement of the M_1- and M_2-lines relative to the N-line (the maximum of which coincided with the position of the Nd^{3+} lines) was significantly greater in $Y_3Al_5O_{12}$ than the splitting of the M-lines determined above relative to the N-line in crystals of $(Y_{1-x}, RE_x)_3Al_5O_{12}$. Along with the appearance of additional lines in the spectra of the garnets of the type considered, there occurred deformation of the shape and broadening of the lines of an N-center (by $\approx$ 1.5 times compared with the lines of the same center in $Y_3Al_5O_{12}$) for L^{2+} concentrations of 8 at.%. In crystals of $(Y_{1-x}Lu_x)Al_5O_{12}$, approximately the same value of line broadening was observed for this transition for similar concentration of Lu^{3+} ions.

Since the M_1 and M_2 spectral lines were only observed in $Y_3Al_5O_{12}-L_3Si_3Al_2O_{12}$ crystals, they can be attributed to centers of composition (7), and we can decide whether they are formed from L^{2+} and Si^{IV} ions simultaneously, or whether these additives appear separately in the M_1- and M_2-centers. In order to do this one can exploit the large difference in the sizes of the Mn^{2+}, Ca^{2+}, and Mg^{2+} ions. This difference is most pronounced between the Mg^{2+} and the Mn^{2+} and Ca^{2+} ions, which in turn have nearly the same ionic radii. If the models for the M_1- and M_2-centers corresponded to a composition given by (7) with n, p $\neq$ 0, a shift of the lines of an AC of a given composition would occur upon substitution of L^{2+} cations. Such a shift should be observed since the environment of an A^{3+} ion is quite sensitive to local deformation of the lattice caused by the presence of an additive in neighboring d-sites. This conclusion follows from an analysis of AC in crystals of mixed rare earth garnets. As is seen from Fig. 13a, the spectra of the crystal $Y_3Al_5O_{12}-Mg_3Al_2Si_3O_{12}$ differ from the spectra of other crystals of this type (Fig. 13b) by the existence of lines corresponding to an M_2-center. However, the M_1-lines in turn (which are present in all three crystals) do not shift when one L^{2+} ion is replaced by another. Thus it can be proposed that an M_1-center has a composition $A^{3+} + pSi^{IV}$ and if the small concentration of Si^{IV} is taken into account, silicon most likely replaces one of the two Al^{3+} ions nearest to Nd^{3+}.

The existence of AC in which L^{2+} ions are present is demonstrated by the fact that deformation of shape and broadening of the N-lines occurs, i.e., the perturbation of the crystal environment of the Nd^{3+} ions is small in this case and, as indicated above, has the same order of magnitude as in the mixed garnet crystals $Y_3Al_5O_{12}-Lu_3Al_5O_{12}$. Consequently, the fact that the L^{2+} ions have a valence different from the Y^{3+} cations does not affect the amount of distortion of the crystal environment of the A^{3+} ion when divalent additives are present in neighboring d-sites of the lattice. On the other hand, for AC of the type $A^{3+} + Si^{IV}$, there is a sizable displacement of the spectral lines relative to the lines of an M-center. This phenomenon could be due to the fact that Si^{IV+} is located at the c-sites of the lattice, and in this case the perturbation of the crystal field near the A^{3+} ion is significantly greater than when there is replacement of the additive in the d-sublattice.

c. In this context it was of interest to investigate $Y_3Al_5O_{12}$ crystals in which Al^{3+} was replaced by a cation with the same valence but different size. This situation occurred in the crystals of $Y_3Al_{5-x}Sc_xO_{12}$ with x = 0-0.5. Lines of two additional centers S_1 and S_2 (Fig. 13c, Fig. 11b), which could be attributed to associates of $A^{3+} + Sc^{3+}$, were observed in the spectra of crystals of this type. In the regions of concentration of Sc^{3+} studied, the intensities of the S_1- and S_2-lines increased steadily with x and no redistribution of intensity between them was observed. This behavior of the S-satellites together with the fact that two types of S-centers occurred simultaneously at extremely low concentrations of Sc^{3+} (0.1 at.%) excludes a model of S-centers as associates composed of different numbers of Sc^{3+} ions occupying only one of the possible Al^{3+} positions (a or c). In this case a redistribution of the intensities of the S_1- and S_2-lines would be observed as the concentration of Sc^{3+} varied, which is not observed experimentally. The reason for the peculiar behavior of the S-centers was successfully explained by studying the defects in the $Y_2Al_5O_{12}$ crystals (Chapter 2).

In the study of the spectra of the crystal $Y_3Al_{5-x}Sc_xO_{12} - A^{3+}$, it was remarked that the shift of the S-center lines relative to N has the same order of magnitude as for the case of M_2- and N-centers in a $Y_3Al_5O_{12} - L_2Si_3Al_2O_{12}$ crystal. Thus it is a general fact for $Y_3Al_5O_{12}$ crystals that strongest perturbation of the crystal environment of an A^{3+} ion, as compared with replacement of yttrium ions, occurs when the supplemental additive is located in the Al-sublattice.

If we take into account that the shifts of the lines of the centers $A^{3+} + Si^{IV}$ and $A^{3+} + Sc^{3+}$ are nearly the same (Fig. 13b, c), it follows that the valence of the ion here also has no essential effect on the structure of the crystal environment of the A^{3+} ion during formation of $A^{3+} + Sc^{3+}$ and $A^{3+} + Si^{IV+}$ associates, and therefore the perturbation introduced by the Sc^{3+} and Si^{IV+} ions is clearly due to local deformation arising from the different sizes of the replacing ions. The greater sensitivity of the symmetry of the environment of A^{3+} ions to perturbations due to replacement of Al^{3+} by another cation can probably be attributed to the fact that this replacement disrupts the structure of the Al–O complex. The O^{2-} ions appearing in this complex are the nearest neighbors of the A^{3+} ion and hence the local perturbation caused by introduction of an additive cation into the Al^{3+} site directly affects the structure of the oxygen environment of the A^{3+} ion. On the basis of the investigations described above of the properties of AC for cases in which the additive is distributed differently in the sites of the garnet lattice, it is possible to explain the large shift of the spectral lines of an M_2-center in the $Y_3Al_5O_{12} - Mg_3Si_3Al_2O_{12}$ crystals (Fig. 13a). Since the size of Mg^{2+} is small enough to permit occupation of the Al^{3+} sublattice sites, it can be hypothesized that an M_2-center forms as a result of such a replacement.

§4. Lasing Properties of $(Y_{1-x}Lu_x)_3Al_5O_{12} - Nd^{3+}$ Crystals

In spite of the comparatively small change in the spectroscopic properties of mixed garnet crystals compared with "simple" garnet crystals, their laser characteristics differ markedly in a number of cases. One such case is the effect observed by us of multitransition generation in $(Y_{1-x}Lu_x)_3Al_5O_{12} - Nd^{3+}$ crystals at 77°K. This phenomenon becomes possible due to broadening of the spectral lines of the transition $^4F_{3/2} \rightarrow {^4I_{11/2}}$ of Nd^{3+} at 77°K as we pass from $Y_3Al_5O_{12}$ to $Lu_3Al_5O_{12}$ in the homologous garnet series. As is seen from Fig. 14d, e, f, the positions of the lines corresponding to transitions B and C exchange places when one garnet is replaced by another. In this connection, the degree of overlapping of the contours of these lines will vary smoothly over wide limits right up to exact resonance as we pass along the homolo-

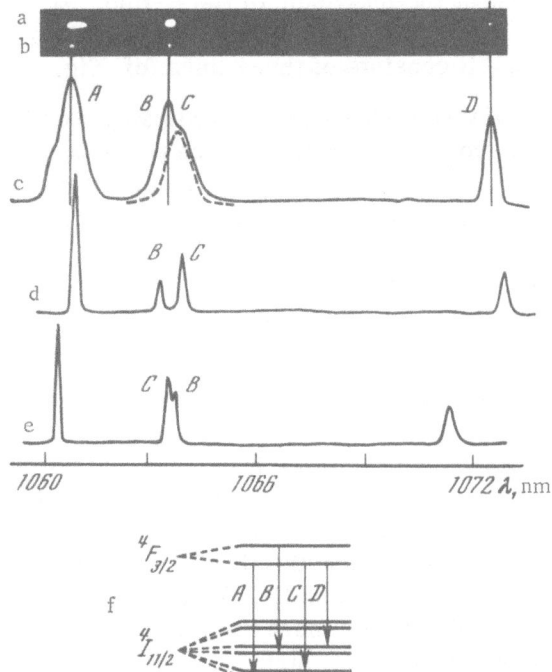

Fig. 14. Induced emission and luminescence spectra of Nd^{3+} in $Y_3Al_5O_{12}$, $Lu_3Al_5O_{12}$, and $(Y_{1-x}Lu_x)_3Al_5O_{12}$ crystals at 77°K. a) Lasing spectrum of $Y_{2.14}Lu_{0.852}Al_5O_{12}$ crystal at pump energy $E_{pump} = 1.1 E_{thr}^{A}$; b) lasing spectrum of the same crystal as in (a) at $E_{pump} = 1.02 E_{thr}^{A}$; c) luminescence spectrum of $Y_{2.14}Lu_{0.852}Nd_{0.003} \cdot Al_5O_{12}$ crystal (dashed line shows the contour of line C at 4.2°K); d) luminescence spectrum of $Y_3Al_5O_{12} - Nd^{3+}$; e) luminescence spectrum of $Lu_3Al_5O_{12} - Nd^{3+}$; f) splitting pattern of the Nd^{3+} levels in garnet.

gous series. Because of this, the relative gain, which is proportional to the intensities of the luminescence lines at the frequencies of the A- or D-line and complex B or C band, will vary as x changes in $(Y_{1-x}Lu_x)_3Al_5O_{12}$, as is illustrated in Fig. 15. It is seen from Fig. 15 that the greatest amplification in $Y_3Al_5O_{12} - Nd^{3+}$ and $Lu_3Al_5O_{12} - Nd^{3+}$ is observed at the A-line frequency, and it is by this transition that lasing should occur. For other values of x in $(Y_{1-x}Lu_x)_3Al_5O_{12} - Nd^{3+}$ a redistribution of intensities is observed, and in a crystal with 80 mol.% $Lu_3Al_5O_{12}$, the maximum gain is observed at the frequency of the complex B, C band (exact resonance of B- and C-transitions).

At a concentration of Lu^{3+} of ≈ 30 at.%, the intensities at the frequencies of B and C become approximately the same. The luminescence spectrum of this crystal is shown in Fig. 14c. The following characteristics were discovered in the investigation of stimulated emission from this sample. At 77°K the spectrum of the induced emission consisted of three lines with wavelengths $\lambda_1 = 10608$, $\lambda_2 = 10636$, and $\lambda_3 = 10726$ Å (Fig. 14a, b) corresponding to the transitions A, BC, and D.

Fig. 15. Dependence of the ratio of luminescence line intensities in the group $^4F_{3/2} \rightarrow {}^4I_{11/2}(I_{BC}/I_A)$ (a) and I_D/I_A (b) on x in $(Y_{1-x}Lu_x)_3Al_5O_{12}$ at 77°K. The designations of the lines are as in Fig. 14. In crystal for which the B and C lines do not overlap, the graph shows I_{BC}/I_A in place of I_C/I_A. The arrow shows the value of x at which the induced emission was studied.

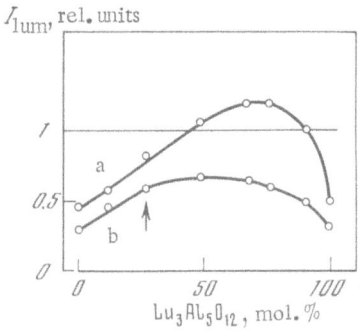

At minimal pump energy (E_{pump}), lasing developed at the maximum of the A-line. At $E_{pump} = 1.02 E_{thr}^{A}$, where E_{thr}^{A} is the threshold value for generation of the A-line, the spectrum consists of two lines (cf. Fig. 14b), and at $E_{pump} = 1.1 E_{thr}^{A}$ it consists of three lines (cf. Fig. 14a).

We consider the factors which lead to generation of three lines with approximately equal thresholds at the four frequencies A, B, C, and D in the group $^4F_{3/2} \rightarrow {}^4I_{11/2}$. The transitions A, C, and D have the same upper level, and B and D have the same lower level. It should first of all be noted that in the present case, multitransition lasing cannot be explained by slow relaxation (w_{21}) between $^4I_{11/2}$ and $^4I_{9/2}$, which was the case in LaF_3-Nd^{3+} crystals [32]. In $Y_3Al_5O_{12} - Nd^{3+}$ crystals, as follows from the results of the measurements performed in [33], the value of w_{21} is equal to $10^7 \sec^{-1}$, which is several orders of magnitude greater than w_{21} in LaF_3.

Multitransition lasing in $(YLu)_3Al_5O_{12}$ becomes possible due to the presence in the host crystal of several species of activator centers. Consider the case where $E_{pump} = 1.02\ E_{thr}^{A}$ (cf. Fig. 14b) when lasing is observed at the maximum of the A-line at frequency ν_A, and at the maximum of the BC line at frequency ν_{BC}. It is seen from a comparison of the lasing and luminescence spectra (cf. Fig. 14b, d) that the frequency ν_{BC} is not at the maximum, but rather lies on the short-wavelength tail of the C-line. If we take into account that when the pump energy is increased, lasing occurs only at the maximum and on the long-wavelength tail of the A-line (cf. Fig. 14a, c), in the case when the character of broadening of the A- and C-lines at frequencies ν_A and ν_{BC} is the same, different groups of centers should for the most part be involved in lasing. The fact that broadening of the A- and C-lines is the same is illustrated in Fig. 12a, which shows the luminescence spectrum of the crystal $(YLu)_3Al_5O_{12}$ at $4.2°K$. It is clear from Fig. 12 that lines of the same types of centers lie at the maxima and on the tails of the A- and C-contours. It can also be shown that lasing by these centers occurs almost independently of one another. First, the rate of energy transfer between the centers in $(YLu)_3Al_5O_{12}$ is small compared to the probability of induced emission (this is indicated by the experimental fact that the A-line becomes strongly broadened with increasing pump energy). Second, different groups of centers cannot be coupled by the emission field on the A and BC transitions in view of the weak overlapping of the lines of the individual groups of centers.* Thus different groups of lines will indeed participate in lasing via the transitions A and C. Hence after the appearance of lasing by the A-transition, the gain k at the frequency ν_{BC} will continue to increase with increasing pump energy, and at $E_{pump} = 1.02 E_{thr}^{A}$, when the gain exceeds the loss coefficient k_{loss}, a second line appears in the lasing spectrum. Lasing develops at the maximum of the D-line when E_{pump} exceeds E_{thr}^{A} by just 10%. Because the width of the luminescence line D is only twice as large as the width of the same line in $Y_3Al_5O_{12}$ (for comparison, we point out that the width of the A-line in $(YLu)_3Al_5O_{12}$ is eight to nine times greater than in $Y_3Al_5O_{12}$), there is extensive overlapping of the lines of almost all types of Nd^{3+} centers in $(YLu)_3Al_5O_{12}$, including the centers not participating in the previously established lasing at frequencies ν_A and ν_{BC}. It is these latter centers which give rise to the increase of the total gain at frequency ν_D as the pump energy increases prior to the time when $k > k_{thr}$.

At $293°K$, lasing develops only at the frequency ν_{BC} of the BC-line. This is due in part to the fact that owing to an increase of the population of the upper Stark component of the level $^4F_{3/2}$, the gain at frequency ν_{BC} is much greater than the gain for the other transitions. In addition, at $293°K$ efficient energy transfer can occur between the various Nd^{3+} centers. Therefore, practically all types of Nd^{3+} centers in $(YLu)_3Al_5O_{12}$ can participate in lasing at the frequency ν_{BC}, and induced emission via the other transitions cannot occur.

---

*If the width of the luminescence line of Nd^{3+} in $Y_3Al_5O_{12}$ is taken as the linewidth of an individual center, then it turns out that there exist approximately eight types of centers within the contours of lines A and C in a $(YLu)_3Al_5O_{12}$ crystal, i.e., approximately the same number of centers which, according to our calculations, is present in $(YLu)_3Al_5O_{12}$.

CHAPTER 2

INVESTIGATION OF LOCAL INHOMOGENEITY OF THE LATTICE OF $RE_3Al_5O_{12}$ GARNETS WITH RARE-EARTH ACTIVATORS

§1. Identification of the Spectral Lines of AC Caused by Various Structural Defects in Garnet Crystals

In the previous chapter we analyzed the mechanism of formation of AC consisting of additive ions in $RE_3Al_5O_{12}$ garnets. The appearance of the AC is due to the formation of associates consisting of the additives introduced into the host crystal. It follows from [5] that such effects can be practically eliminated by reducing the concentration of the additive ions. However, in garnet crystals with TR^{3+} additive ions, a fine structure of the spectral lines was recorded even for cases of limitingly small concentrations of activators, and the relative intensities of the satellites did not depend on the concentration of the additive ions. This fact indicates that there exist defects in garnet crystals whose appearance is not related to the introduction of additive ions. At the same time, these additives permit us to record the existence of defects of this kind, since they serve to indicate local distortions of the lattice. In order to determine the reason for the defects in garnet crystals, it was necessary to analyze all the known types of lattice distortions occurring most commonly in crystals, viz. dislocations, incorporation of material from the container or uncontrolled impurities, and the presence of thermal distortions. In addition, it was necessary to take into account the possibility of lattice distortion merely by introducing a small number of indicator ions if the dimensions of the latter do not correspond to the ionic radius of the RE^{3+} cation, e.g., Nd^{3+} in $RE_3Al_5O_{12}$.

Clearly, equilibrium proper defects (defects in the sense of Schottky and Frenkel) need not be considered, since ordinarily their concentrations are negligibly small, 10^{15}-10^{16} cm^{-3} [31]. By exploiting the fact that their concentrations can be redistributed by suitably changing the conditions of synthesis, it is possible to study the nature of the local disruptions of lattice periodicity considered above. Therefore, if the fine structure of the spectral lines of the indicator ions is caused by any defects of the above types, this can be established by following the correlation among the intensities of certain satellites and the change of the concentrations of the corresponding types of defects.

We consider the character of the inhomogeneous splitting of the spectral lines of A^{3+} ions in $RE_3Al_5O_{12}$ crystals. Figure 16a, b shows the absorption and luminescence spectra of Nd^{3+} (1 at.%) in the group $^4I_{9/2} \rightleftharpoons {}^4F_{3/2}$ in $Y_3Al_5O_{12}$ at 4.2°K. We can conclude from an analysis of the spectra that one of the observable satellites of the principal N-line, viz. the M-satellite, can be attributed to paired $Nd^{3+} + Nd^{3+}$ associates for the following two reasons: First, the dependence of the relative intensity of the M-line, $j_M = \sigma_M / (\sigma_M + \sigma_N)$ (the integrated absorption coefficient) on the concentration of Nd^{3+} in the crystal coincides with the behavior of the change of the relative concentration of the pair centers calculated by Eq. (9) (Fig. 16c); and second, the M-line is quenched in the luminescence spectrum, which is also a criterion for associates of Nd^{3+} ions [34] (Fig. 16b). It should be noted that the lines of the pair or more complex associates of A^{3+} ions could be distinguished only in the case when the size of the indicator ions differed greatly from the size of the RE^{3+} cation. If this difference was small (when the A^{3+} ions were Er^{3+} and Yb^{3+}) it was not possible to observe the lines of the corresponding centers due to the extremely small displacement of the lines of the associates relative to the lines of isolated N-center. This effect was discussed in the previous chapter. A distinctive feature of the other observable satellites in the spectra of the ions is their nondependence on the concentration (Fig. 16a, c). Hence these particular centers, which we have called $\mathscr{P}$-centers, are not involved with the formation of associates of A^{3+} ions. In order to ascertain their nature, a

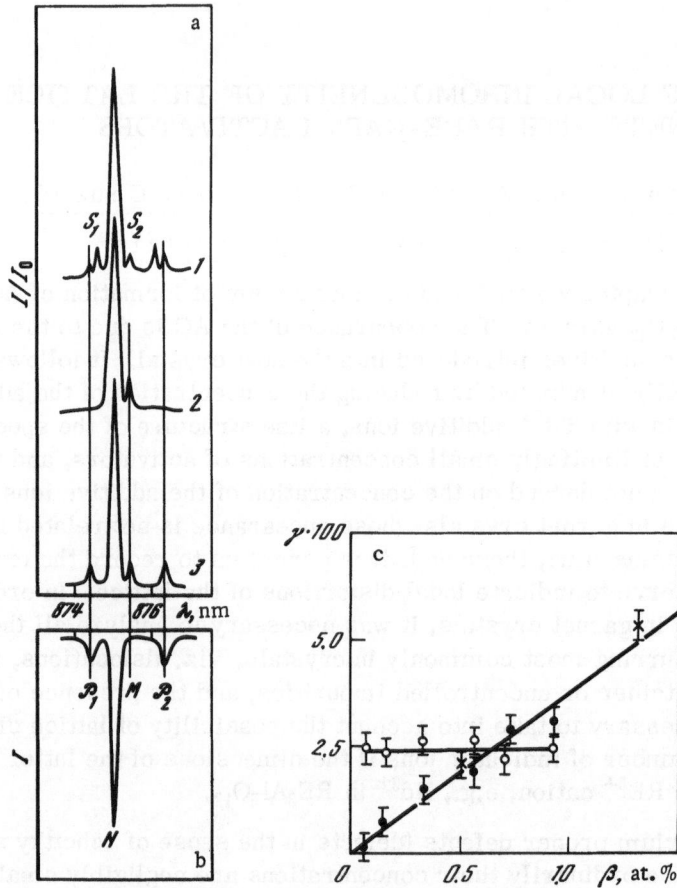

Fig. 16. Absorption spectrum Nd^{3+} for the $^4I_{9/2} \to {}^4F_{3/2}$ transition at 4.2°K in the crystals (a): $Y_3Al_5O_{12} - Sc^{3+}$ (10 at.%), crystal grown from a melt (1), $Y_3Al_5O_{12}$, crystal grown from solution in a melt (2), $Y_3Al_5O_{12}$, crystal grown from a melt (3); luminescence spectrum of Nd^{3+} for the transition $^4F_{3/2} \to {}^4I_{9/2}$ in a $Y_3Al_5O_{12}$ crystal obtained from a melt (b); dependence of the integrated absorption coefficient of the spectral lines denoted by M, N, $\mathscr{P}_{1,2}$, on the concentration of Nd^{3+} in the crystal (c). The sloped line corresponds to the calculated dependence of the change of concentration of $Nd^{3+} + Nd^{3+}$ pairs.

series of experiments was performed in which the concentration of structural defects of one of the types considered above was varied systematically by changing the conditions of synthesis.

We studied crystals grown using different techniques, viz., the Czochralski method using iridium crucibles in atmosphere of nitrogen, of argon, and in a vacuum, the method of oriented crystallization from a molybdenum container in an argon atmosphere, the method of crucible-free zone melting, and by solution in a melt. The application of such diverse methods of synthesis made it possible to change the number of defects such as dislocations in the crystals (the density of dislocations in our crystals changed from 10^2 to 10^6 cm^{-2}). The investigations showed that in all the crystals obtained by various techniques from a melt and containing essentially different numbers of dislocations, the quantity $j_{\mathscr{P}} = \dfrac{\sigma_{\mathscr{P}}}{\sigma_N + \sigma_{\mathscr{P}}}$ remains the same. This fact permits us to eliminate dislocations and microinclusion from the container from the possible factors responsible for the appearance of the $\mathscr{P}$-centers. The conclusion concerning microinclusions is valid because the methods for obtaining single crystals enumerated above involve the use of the most diverse container materials (iridium, molybdenum, without crucible). The presence in the crystal of uncontrolled impurities introduced by the initial components also cannot be a reason for the appearance of the $\mathscr{P}$-centers of A^{3+} ions, since the values of $j_{\mathscr{P}}$ did not change in crystals grown from materials with varying contents of foreign impurities. The purest of the components used by us contained foreign impurities in amounts not exceeding 10^{-5}-$10^{-6}\%$. Spectroscopic studies of the garnet crystals before and after prolonged annealing at high temperatures did not reveal any change in the relative intensities of the $\mathscr{P}$-lines. Consequently, thermal distortions play no role in the formation of $\mathscr{P}$-centers. It should also be noted that change of symmetry of the unit cell due to the introduction of A^{3+} ions with dimensions differing greatly from the dimensions of the RE^{3+} cations cannot be a reason for the appearance of $\mathscr{P}$-centers. In order to study defects of this type, we used as A^{3+} ions the series of TR^{3+} activators Nd^{3+}, Er^{3+}, Eu^{3+}, Yb^{3+}. As was already pointed out in Chap. 1, Sec. 1, these ions are such that the ratios of their dimensions compared to the ionic radius of the RE^{3+} cation differ maximally (this ratio changes as we move along the $RE_3Al_5O_{12}$ series). Figure 17a shows graphs of the function $j_{\mathscr{P}}(z)$ (z is the atomic number of the RE^{3+} cation in the lanthanide series) for all four A^{3+} ions. It is clear from this figure that all the curves $j_{\mathscr{P}}(z)$ coincide, and hence the introduction of indicator ions indeed does not result in the appearance of $\mathscr{P}$-centers. The same conclusion is supported by a study of the spectra of Er^{3+} in $Re_3Al_5O_{12}$ and Yb^{3+} in $Yb_3Al_5O_{12}$, where the ions indicated are not additives, but themselves form the RE^{3+} sublattice of $RE_3Al_5O_{12}$. Lines from $\mathscr{P}$-centers were also recorded in these crystals. Thus, it follows from the above experimental results that specific structural defects are present in the lattice of garnets obtained from a melt which remove the structural degeneracy of the d-sites of the lattice and cause the appearance of activator $\mathscr{P}$-centers of TR^{3+} ion additives.

§2. Effect of High-Temperature Structural Disordering in Garnet Crystals

The above results indicate that the lattices of garnet crystals obtained from a melt contain a certain number of inequivalent d-sites. In contrast, such an inequivalence was not ob-

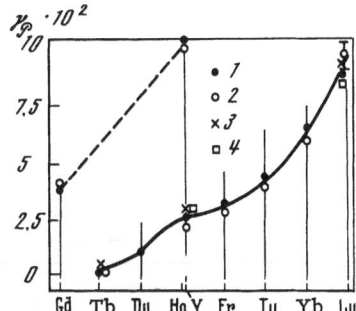

Fig. 17. Dependence of the quantity $\gamma \mathscr{P} j_{\mathscr{P}} = \sigma_{\mathscr{P}}/(\sigma_{\mathscr{P}} + \sigma_N)$ on the atomic number of the RE^{3+} cation in $RE_3Al_5O_{12}$ (a) and $RE_3Ga_5O_{12}$ crystals (b) for the following indicator ions: 1) Er^{3+}; 2) Nd^{3+}; 3) Eu^{3+}; 4) Yb^{3+}.

served in garnet crystals obtained from solution in a melt. The absorption spectrum of a $Y_3Al_5O_{12}-Nd^{3+}$ crystal obtained by this method (Fig. 16a) contains no $\mathscr{P}$-lines. We may thus conclude that the structure of the garnet lattice differs from the cases of synthesis at high, 1960°C, and low, 1050°C, temperatures. Two hypotheses may be considered here. According to the first hypothesis, there exists a "high-temperature" modification of garnets for which the symmetry of the unit cell differs slightly from cubic and which gives rise to the existence of inequivalent lattice sites; when the crystal is cooled, the structure of this modification is preserved. The second hypothesis is that at high temperatures, local inhomogeneities of microscopic dimensions (of the order of the lattice constant) form in the crystal which have a definite structure differing from that of the bulk of the crystal (this possibility was pointed out in [35] on the basis of the results of an investigation of $Y_3Al_5O_{12}$ using an electron microscope).

The results of x-ray analysis lend support to the second hypothesis. These results revealed a complete agreement of the diffraction patterns of crystals obtained from a melt and from solution in a melt, but the existence of other phases was detected in the "high-temperature" garnet crystals.

In order to ascertain the nature of the local lattice distortions leading to the formation of $\mathscr{P}$-centers, we studied the changes of the lattice structure which can affect the concentration of $\mathscr{P}$-centers.

It was shown above that the relative intensities of the lines of the $\mathscr{P}$-centers vary as we move along the $Re_3Al_5O_{12}$ series (Fig. 17a). It follows that in $Yb_3Al_5O_{12}$ the concentration of $\mathscr{P}$-centers is practically zero, while the concentration reaches a maximal value in $Lu_3Al_5O_{12}$. As we move along the $RE_3Al_5O_{12}$ series, the number of $\mathscr{P}$-satellites remains the same, and depending on the transition observed, two or three $\mathscr{P}$-lines of equal intensity were observed (Figs. 18, 19). The position of the $\mathscr{P}$-lines relative to N does not change either as we pass along the $RE_3Al_5O_{12}$ series. Since the symmetry of the unit cell is the same for each member of the $RE_3Al_5O_{12}$ series [9], the observed behavior may be attributed to a variation of the dimensions ψ of the unit cell, which in this case are determined by the dimensions of the RE^{3+} cations.

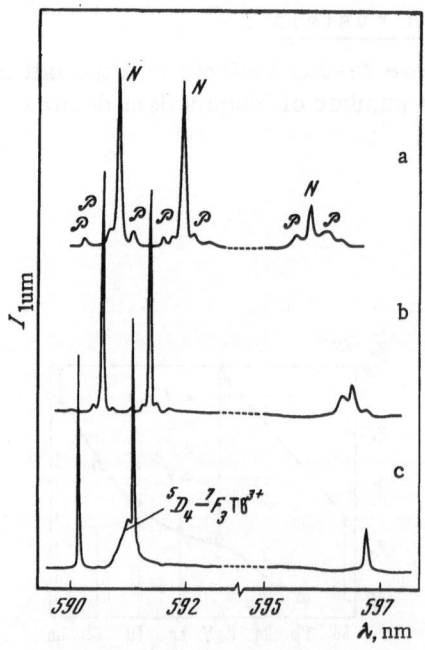

Fig. 18. Luminescence spectra for Eu^{3+} (0.1 at.%) in the group $^5D_0 \rightarrow {}^7F_1$ at 4.2°K in the crystals a) $Lu_3Al_5O_{12}$; b) $Y_3Al_5O_{12}$; c) $Tb_3Al_5O_{12}$.

The concentration of $\mathscr{P}$-centers in $Y_3Al_5O_{12}$ could also be changed by introducing other cations into the host crystal in place of Y^{3+}. The studies were carried out with crystals of $(Y_{1-x}Lu_x)_3Al_5O_{12}$ and $(Y_{1-x}Gd_x)_3Al_5O_{12}$. As the concentration of the Lu^{3+} ions increased, an increase was observed in the intensities of the $\mathscr{P}$-lines, and conversely, when the large cations Gd^{3+} were introduced, their intensity decreased. This is illustrated in Fig. 20, which shows the curves $i_{\mathscr{P}}$ as a function of the dimensions of the unit cell, the lattice constant ψ of which is uniquely determined by the concentration of Lu^{3+} or Gd^{3+} in $Y_3Al_5O_{12}$. Figure 20 also shows the the function $i_{\mathscr{P}}(\psi)$ for the $RE_3Al_5O_{12}$ series, which as is seen from the figure behaves like the composite curve $(Y_{1-x}Lu_x)_3Al_5O_{12}$ and $(Y_{1-x}Gd_x)_3Al_5O_{12}$.

It should be remarked that the change in the concentration of $\mathscr{P}$-centers was monitored by us in terms of the change of $i_{\mathscr{P}}$, the relative intensity of one of the $\mathscr{P}$-satellites. This can be done if the probabilities of transitions in the $\mathscr{P}$- and N-lines are nearly the same. We show that this conditions is satisfied in our case.

On the basis of the extremely small displacement of the $\mathscr{P}$-satellites relative to the N-line (1-5 cm^{-1}) we may conclude that the perturbation of the crystal field near an A^{3+} ion in the presence of a nearby defect forming a $\mathscr{P}$-center is extremely slight, and hence it should not change the probability of transition in a $\mathscr{P}$-line relative to N. If such a change nevertheless occurred, it would cause a divergence among the values of $i_{\mathscr{P}}$, measured for different groups of lines, and a large difference should be observed for the case of electric and magnetic dipole transitions. The latter (cf. [36]) are practically insensitive to perturbations of the crystal field. Since, however, the experimental value of $i_{\mathscr{P}}$, measured for different transitions, including magnetic-dipole transitions, had the same value for different A^{3+} ions, our assumption that the transitions probabilities in the N- and $\mathscr{P}$-lines are nearly the same is legitimate, and the quantity $i_{\mathscr{P}}$ corresponds to the relative concentration of one of the $\mathscr{P}$-centers.

It can be concluded from the above results that first, the concentration of $\mathscr{P}$-centers is determined by the size of the cations making up the RE^{3+} sublattice in $RE_3Al_5O_{12}$.

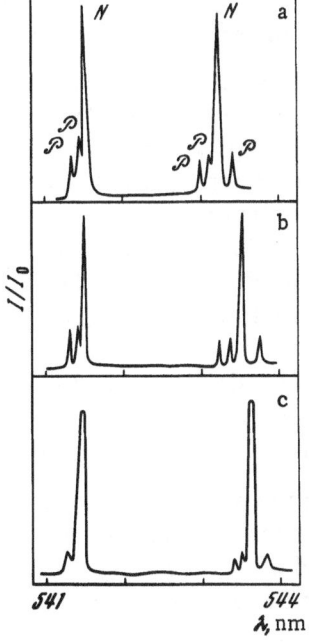

Fig. 19. Absorption spectra of Er^{3+} (0.1 at.%) in the group $^4I_{15/2} \rightarrow \, ^4S_{3/2}$ at 4.2°K in the crystals a) $Lu_3Al_5O_{12}$; b) $Y_3Al_5O_{12}$; c) $Dy_3Al_5O_{12}$.

Fig. 20. Dependence of the quantity $\gamma_{\mathscr{P}}\, i_{\mathscr{P}} = \sigma_{\mathscr{P}}/(\sigma_{\mathscr{P}} + \sigma_N)$ on the dimensions (ψ) of the unit cell for the following compounds: 1) homologous series of garnets $RE_3Al_5 \cdot O_{12}$; 2) series of solid solutions $(Y_{1-x}Yu_x)_3Al_5O_{12}$, x varying continuously; 3) series of solid solutions $Y_3Al_5O_{12} - Gd_3Al_5O_{12}$ to concentration of 20 mol.%; 4) crystals of $Y_3Al_5O_{12} - Sc^{3+}$ (0.5–8 at.%); 5) crystals of $Lu_3Al_5O_{12} - Sc^{3+}$ (0.5–8 at.%).

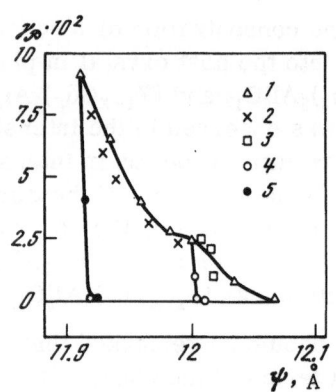

Second, the concentration of $\mathscr{P}$-centers could be decreased to practically zero by changing the structure of the Al-sublattice. Figure 20 shows the curves $j(\psi)$ for crystals of $Y_3Al_5O_{12}$ and $Lu_3Al_5O_{12}$, in which Sc^{3+} cations were introduced in place of Al^{3+}. It is seen from Fig. 20 that $j(\psi)$ drops abruptly for even a very slight increase of the lattice constant, which corresponds to a concentration of Sc^{3+} of ≈ 2 at.%. Such an efficient effect of Sc^{3+} on the concentration of $\mathscr{P}$-centers suggested that the defects responsible for the appearance of $\mathscr{P}$-centers are associated with a disruption of the structure of the aluminum sublattice.

The incorporation of Sc^{3+} in place of Al^{3+} leads to a change of the original symmetry of the lattice d-site, with the result (as follows from Chap. 1) that associates $A^{3+} + Sc^{3+}$ and S_1- and S_2-centers appear.

A comparison of the spectra of the A^{3+} ions in $RE_3Al_5O_{12}$ with the spectra of the same ions in crystals with Sc^{3+} reveals that the number of S-centers corresponds exactly to the number of $\mathscr{P}$-lines for a given transition.

The nature of the position of the $\mathscr{P}$- and S-lines relative to the N-lines is also preserved. This is illustrated in Fig. 16a (curve 1) for the crystal $Y_3Al_5O_{12} - Sc^{3+} - Nd^{3+}$ and the transition $^4I_{9/2} \rightarrow \,^4F_{3/2}$ of Nd^{3+}.

This suggested that the local deformation responsible for the appearance of $\mathscr{P}$-centers is similar to the deformation occurring when Al^{3+} is replaced by a larger cation.

The role of this cation can be played by RE^{3+} ions forming the d-sublattice in $RE_3Al_5O_{12}$. In this case we should observe a departure from strict stoichiometry of the garnet in favor of excess RE_2O_3. In the series of papers [37–39] the possibility of introducing an excess of RE^{3+} ions in place of the octahedral Al^{3+} (d-sites) was discussed. Evidence that this is possible was provided by experiments in which the lattice constant was measured to high precision as a function of the crystal composition. These experiments were carried out with gallium crystals, which have a comparatively large region of homogeneity, and showed that the dimensions ψ increase with increasing RE_2O_3 excess in the crystal. By analogy with gallium garnets, the same conclusions were extended to aluminum garnets since an increase in the lattice constant as compared with "stoichiometric" crystals [37] was also recorded in $Y_3Al_5O_{12}$ in crystals obtained by the Czochralski method. By "stoichiometric" crystals, we mean polycrystals obtained by solid-phase reaction, and single crystals grown from solution in a melt.

It may be noted here that this experimental fact is in accordance with the features of the appearance of $\mathscr{P}$-centers recorded by us in crystals obtained from a melt, and with the absence of $\mathscr{P}$-centers in crystals prepared by the method of solution in a melt. This result did not indicate with complete certainty that the reason for the formation of $\mathscr{P}$-centers is the superstoichiometric excess of RE^{3+} in place of Al^{3+}.

The best proof that this hypothesis is correct might have been obtained from studying the correlations between the concentration of $\mathscr{P}$-centers and the composition of $RE_3Al_5O_{12}$. Such studies were carried out by us for aluminum garnet crystals, but did not give unambiguous results because it was not possible to vary the composition of the melt over wide limits and still maintain the presence of only one phase in the crystals.

Therefore, in [40, 41] the cause of $\mathscr{P}$-center formation was attributed to defects which form, according to [31], "orientational disordering." In garnet crystals, this type of defect could be associated with bending of the octahedral or tetrahedral Al—O bonds at high temperature.

By synthesizing single crystals of gallium $RE_3Ga_5O_{12}$ garnets, which in terms of their crystal chemistry are analogous to aluminum garnets, we were able to clarify the reason for the formation of $\mathscr{P}$-centers.

It was discovered in the investigation of the spectra of Nd^{3+} and Er^{3+} in $Gd_3Ga_5O_{12}$ and $Y_3Ga_5O_{12}$ (the melt corresponded to the stoichiometric composition) that $\mathscr{P}$-centers are also observed in this series of crystals (Fig. 21). Like aluminum garnets, an increase in the intensities of the $\mathscr{P}$-lines was observed with increasing size of the RE^{3+} cations, while at the same time the relative intensity of the $\mathscr{P}$-lines in the $RE_3Ga_5O_{12}$ series was significantly greater than in $RE_3Al_5O_{12}$ (cf. Fig. 17a, b).

Since in contrast to aluminum garnets, the region of homogeneous composition is much larger, it was possible to perform experiments in which the composition of the crystal was varied over wider limits.

In our experiments we studied a series of yttrium—gallium garnet crystals grown from a melt in which the concentration of Y_2O_3 varied from 5% insufficiency to 15% excess compared to the stoichiometric composition. The intensities of the $\mathscr{P}$-lines in crystals with insufficient Y_2O_3 turned out to be the same as for the "stoichiometric crystals" while at the same time, in crystals with excess Y_2O_3 an increase of the intensity of $\mathscr{P}$-satellites was observed as the concentration of Y_2O_3 increased in the crystal (cf. Fig. 21).

Thus, the experimental results for the gallium garnets made it possible to unambiguously attribute the appearance of $\mathscr{P}$-centers and inhomogeneous line splitting of RE^{3+} ion additives

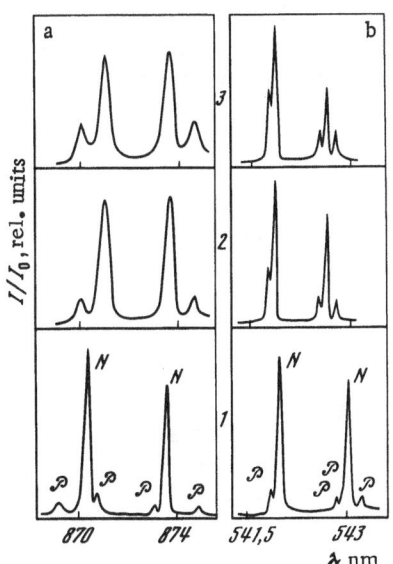

Fig. 21. Absorption spectra of Nd^{3+} in the group $^4I_{9/2} \rightarrow {}^4F_{3/2}$ (a) and of Er^{3+} in the group $^4I_{15/2} \rightarrow {}^4S_{3/2}$ (b) at 4.2°K in crystals grown from melts: 1) $Gd_3Ga_5O_{12}$; 2) $Y_3Ga_5O_{12}$; 3) $Y_3(Ga_{1.8}Y_{0.2})Ga_3O_{12}$.

in gallium, and hence also in aluminum garnets obtained from a melt to the presence of a superstoichiometric excess of RE^{3+} ions at the Al^{3+} positions. In addition, the spectroscopic investigations carried out confirmed the possibility of replacing Al^{3+} by rare-earth ions in aluminum garnets, which in contrast to gallium garnets required additional argument [37].

The fact that incorporation of RE^{3+} ions in place of Al^{3+} occurs only at high temperature can be completely explained by the theory of solutions and statistical thermodynamics [31].

Indeed, the process of incorporation of RE^{3+} ions in place of Al^{3+} must, owing to the large size of the RE^{3+} ions, be accompanied by significant energy losses, which can be compensated for by increased configurational and vibrational entropy. Since the role of entropy in the Gibbs free energy becomes significant at high temperature, incorporation of an RE^{3+} ion into the Al^{3+} lattice sites can only occur in crystals synthesized from a melt.

We now consider a possible model for the $\mathcal{P}$-centers. Starting from the fact that RE^{3+} ions are incorporated in place of Al^{3+} (where on the basis of [37] octahedral Al^{3+}, i.e., a-sites, must be considered here), the formation of $\mathcal{P}$-centers can be related to the appearance of an associate of A^{3+} and RE^{3+} ions in an a-site of the lattice (RE_a^{3+}). In this case, perturbation of the environment of the A^{3+} ion should occur as compared with the isolated N-center model, because the sizes of the RE^{3+} and Al^{3+} ions are different. Since preliminary estimates of the concentration of RE_a^{3+} ions from data of [37] give a value of several percent, it follows from statistical considerations based on equiprobable distribution of the ions that the binary associate of the type $A^{3+} + RE_a^{3+}$ should predominate. It is appropriate to consider here some facts which at first glance seem contradictory. If the $\mathcal{P}$-centers are caused by the presence of the pairs $A^{3+} + RE_a^{3+}$ then by analogy with the formation of binary centers of the type $A^{3+} + RE_a^{3+}$ (Chap. 1), we should have observed at most one $\mathcal{P}$-satellite instead of the three which are actually recorded in the spectra of some transitions (cf. Figs. 18, 19). This contradiction is easily resolved if we bear in mind the fact that the intensities of the $\mathcal{P}$-lines are identical, which indicated that the concentrations of the three types of $\mathcal{P}$-centers are also equal. Such a situation is possible if the incorporation of one RE^{3+} ion into a lattice d-site perturbs the environments of the three closest lattice d-sites in different ways. It follows from the structure of the garnet lattice that this actually can occur.

Figure 22 shows a possible model for a $\mathcal{P}$-center. It is clear from the figure that octahedral Al^{3+} is situated at different distances from the three lattice d-sites. If the perturbation of the crystal field for these sites was determined solely by the interaction among the ions located at the a- and d-sites, then we would have a threefold degeneracy of the $\mathcal{P}$-centers and only one $\mathcal{P}$-satellite would be observed in the spectra. Therefore, the most likely result of incorporation of an RE_a^{3+} ion is to deform the oxygen octahedron surrounding an a-site, as a result of which the O^{2-} ions which appear as neighbors of both the a- and d-sites are shifted, so that nonidentical crystal fields are formed for the three ions located at each of the three lattice d-sites.

An analogous explanation can be given for the interpretation of the previously obtained results concerning the existence of several satellites with equal intensity and belonging to the $A^{3+} + Sc^{3+}$ centers.

Fig. 22. Model of an $A^{3+} + RE^{3+}$ center in $RE_3(RE\ Al_{1-x})_2Al_3O_{12}$ crystals.

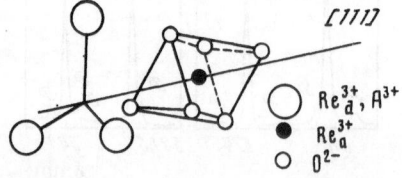

Taking into account the fact that all the garnet crystals obtained from a melt contain an excess of RE^{3+}, it is not altogether correct to write their formula as $RE_3Al_5O_{12}$. The actual formula for these garnets is $RE_3Al_3RE_xAl_{2-x}O_{12}$. However, bearing in mind that the size of the deviation from the stoichiometric composition by [37] does not exceed 2-3%, we will for convenience retain the old notation in what follows and indicate where necessary the means by which the crystal was obtained.

It should be noted that the model proposed by us of a $\mathscr{P}$-center as a binary associate $A^{3+} + RE_a^{3+}$ makes it possible to determine the concentration of RE_a^{3+} ions in terms of the line intensities of the $\mathscr{P}$-centers. As shown above, the quantity $j_{\mathscr{P}}$ corresponds to the relative concentration of one of the three $\mathscr{P}$-centers, i.e., to the concentration of the A^{3+} ions appearing in the binary associates $A^{3+} + RE_a^{3+}$. Once this concentration is known, the concentration of the RE^{3+} ions can be found from Eq. (9), which can be easily transformed for the case of $\mathscr{P}$-centers. The calculations we carried out gave a value of 1.5% for the concentration of RE_a^{3+} ions in aluminum−yttrium garnet, in complete agreement with the value found in [37] from the results of x-ray analysis.

Since the quantity $j_{\mathscr{P}}$ is related to the concentration of the RE^{3+} ions in the aluminum sublattice, the change in $j_{\mathscr{P}}$ as we pass along the series of garnets $RE_3Al_5O_{12}$ (which is shown in Fig. 17) illustrates the role played by the sizes of the RE^{3+} ions on the probability of replacement of Al^{3+} by RE^{3+}. It is clear from Fig. 17 that as the size of the RE^{3+} cation decreases (z increases), the concentration of RE^{3+} ions in the lattice a-sites increases. This behavior is completely explainable from the viewpoint of thermodynamics, since as the size of the RE^{3+} ions decreases, energy losses associated with forces of elastic character resulting from the replacement of Al^{3+} by RE^{3+} are reduced.

It is also seen from Fig. 17 that it is not just the sizes of the ions that play a role in the effect considered. Thus, in $Y_3Al_5O_{12}$ and $Lu_3Al_5O_{12}$, a definite jump of $j_{\mathscr{P}}(z)$ toward larger values is observed. This indicates that ions with a spherical electron shell (Y^{3+}, Lu^{3+}) have a greater probability of entering the Al^{3+} sites than the other rare-earth ions.

§ 3. Mutual Effect of Rare-Earth Ions on the Character

of Their Location at Inequivalent Sites of the Crystal Lattice

in Aluminum Garnet Single Crystals

The investigations considered above of replacement of octahedral Al^{3+} by rare-earth ions were for single-component garnets. The concentrations of indicator ions in these crystals were

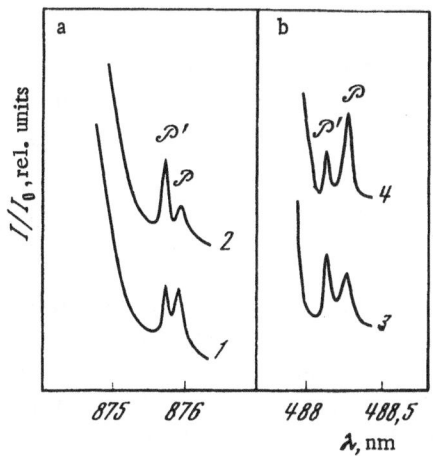

Fig. 23. Change of intensities of the lines of $A_d^{3+} + Y_a^{3+}$ ($\mathscr{P}$) and $A_d^{3+} + Lu_a^{3+}(Yb_a^{3+})$ ($\mathscr{P}'$) centers for the cases: a) $A^{3+} = Nd^{3+}$ for the transition $^4I_{9/2} \rightarrow {}^4F_{3/2}$ at 4.2°K; b) $A^{3+} = Er^{3+}$ for the transition $^4I_{15/2} \rightarrow {}^4F_{7/2}$ at 77°K in which the composition of the mixed garnets was varied: 1) $Y_3Al_5O_{12} - Lu^{3+}$ (7 at.%); 2) $Y_3Al_5O_{12} - Lu^{3+}$ (20 at.%); 3) $Y_3Al_5O_{12} - Lu^{3+}$ (10 at.%); 4) $Y_3Al_5O_{12} - Yb^{3+}$ (10 at.%).

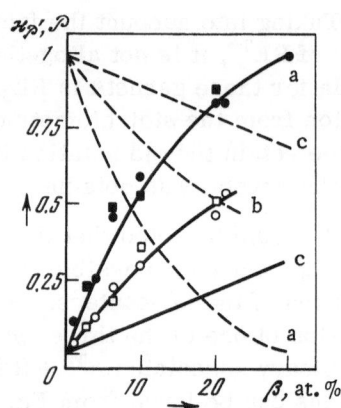

Fig. 24. Relative intensities of the lines of the $\mathscr{P}$ ($\varkappa_{\mathscr{P}}$) centers (dashed lines) and $\mathscr{P}'$ ($\varkappa_{\mathscr{P}'}$) (solid lines) as function of the concentration of Lu(Yb) in the lattice d-sites: a) $(Y_{1-x}Lu_x)_3Al_5O_{12}$; b) $(Y_{1-x}Yb_x)_3 \cdot Al_5O_{12}$; c) calculated dependences of $\varkappa_{\mathscr{P}}$ and $\varkappa_{\mathscr{P}'}$ in the case of proportional change of the concentration of Y_a^{3+} and Yb_a^{3+} (Lu_a^{3+}) ions with increasing Lu(Yb) concentration at the lattice d-sites.

too small to have any significant effect on the process of incorporation of rare-earth ions into the Al sublattice.

On the other hand, in the case of mixed garnets, several species of rare-earth ions are present in Al^{3+} sites, and hence interactions between them are not ruled out.

The effect of interaction of ions located at the Al^{3+} lattice sites was experimentally detected by us in mixed rare-earth garnet crystals.

Figure 23 shows the absorption spectra of the crystals $(Y_{1-x}Lu_x)_3Al_5O_{12}$ and $(Y_{1-x}Yb_x)Al_5O_{12}$ $A^{3+} = Nd^{3+}$, Er^{3+} for transitions of the A^{3+} ions for which the lines of both the centers $A^{3+} + Y_a^{3+}$ ($\mathscr{P}$) and $A^{3+} + Lu_a^{3+}$ (Yb_a^{3+}) ($\mathscr{P}$)' were clearly recorded. As always, in our experiments we used several types of indicator ions in order to take into account any possible effect of them on the processes studied.

Figure 24 shows the dependence of the relative integrated intensities of the absorption lines, $\varkappa_{\mathscr{P}} = \dfrac{\sigma_{\mathscr{P}}}{\sigma_{\mathscr{P}} + \sigma_{\mathscr{P}'}}$ and $\varkappa_{\mathscr{P}'} = \dfrac{\sigma_{\mathscr{P}'}}{\sigma_{\mathscr{P}'} + \sigma_{\mathscr{P}'}}$, of these centers as a function of the concentration of Lu(Yb) (β) in the crystal.

It is clear from Fig. 24 that as β increases, the concentration of $A^{3+} + Y_a^{3+}$ centers decreases much more rapidly than can be explained by the decrease in the concentration of Y^{3+} in the crystal (straight line). For crystals containing Lu^{3+} this effect is much more pronounced than for crystals with Yb^{3+}.

Consequently the Lu^{3+} ions, whose sizes are more suitable for the d-sites of the lattice, "squeeze out" the larger Y^{3+} ions from the a-sublattice.

Let us analyze possible reasons for these experimental results. One such reason might be the elastic interaction between distinct rare-earth ions caused by the fact that their sizes are much larger than the ionic radius of Al^{3+}, which should cause repulsion of the rare-earth ions.

As has been shown above, this type of interaction in aluminum garnets has an extremely small effective radius and therefore clearly cannot have a significant effect on the concentration of ions introduced into the crystal.

The following explanation of the phenomenon can be suggested. During synthesis, rare-earth ions of various types enter the a-sites independently of one another. In a mixed garnet, e.g., in $(Y_{1-x}Lu_x)_3Al_5O_{12}$, we have two types of ions in each of the a- and d-sites, Y_d^{3+}, Lu_d^{3+}, and Y_a^{3+}, Lu_a^{3+}. Since Lu^{3+} has a size better suited for the a-sites than Y^{3+}, it is then possible for exchange to occur — the Y^{3+} ion shifts to the position of the Lu_d^{3+}, and conversely, Lu^{3+} moves from the d-sublattice into the position of Y_a^{3+}. This process is energetically favorable, since larger cations at the Al^{3+} sites are replaced by smaller ions.

The effect we detected of interaction between rare-earth ions with different dimensions on their concentration in the a-sublattice can be explained by many factors related to the change of properties of the garnet crystals containing several TR^{3+} activators, in particular, the role of Lu^{3+} in improving the quality of laser crystals. The point is that incorporation of large RE^{3+} ions at the Al^{3+} sites in $RE_3Al_5O_{12}$ crystals obtained from a melt is unfavorable since it can give rise to internal stresses in the crystal. The introduction of Lu^{3+} and associated effect of "squeezing out" of Y^{3+} ions from the a-sites leads to a situation in which the role of such stresses is lessened owing to replacement of a larger Y^{3+} ion by an Lu^{3+} ion with dimensions better suited for a lattice a-site. It is seen from Fig. 24 that the squeezing out effect is very important, 10-20 at.% Lu^{3+} being required for a significant reduction of the Y^{3+} ion concentration in the crystal. This is the same amount of Lu^{3+} as was introduced into the garnet laser crystals with Nd^{3+} in [25].

CHAPTER 3

STUDY OF THE CHARACTER OF THE INTERACTION IN ASSOCIATES OF RARE-EARTH IONS IN A GARNET HOST CRYSTAL

§1. Conditions under Which Short-Range Interaction in a System of TR^{3+} Ions Is Observed

The study of the processes of formation of associates of TR^{3+} ions in garnet crystals (cf. Chapter 1) makes it possible to investigate the question concerning the mechanism of interaction between rare-earth ions at very short distances leading to radiationless transfer of excitation energy. In [42-44] data are cited from experimental and theoretical investigations of long-range multipole mechanisms of energy transfer. At the same time, there are very few experimental studies concerned with the observation of short-range mechanisms of radiationless energy transfer in a system of rare-earth ions, which according to [45] should predominate when multipole transitions in the donor and acceptor are forbidden.

The theory stipulates the existence of transfer mechanisms which are effective only at very small distances between ions [45-47]. Thus in [45] an exchange mechanism of transfer caused by overlapping of the electron clouds of the interacting ions is considered.

The rare-earth ions are peculiar in that the f-shells lie deep in the interior of the atom so that overlapping of the electron clouds of the rare-earth ions is extremely small, even when the TR^{3+} ions are nearest neighbors in a crystal lattice. According to an estimate in [48], the overlap integral for TR^{3+} ions is approximately 10^3-10^5 times smaller than for ions in the iron group.

For this reason the question of the existence of an exchange mechanism for transfer in a system of TR^{3+} ions required experimental confirmation.

In [49, 50] experiments were performed which provided evidence for an exchange mechanism of transfer. In [49], the evidence came from the experimental fact that the quenching efficiency of the donor (Eu^{3+}, Tb^{3+}) depends on the potential number of acceptor sites in the innermost coordination sphere; in [50], the evidence came from an analysis of the selection rules in the acceptor.

It should be noted that if it were possible to detect an abrupt decrease of the probability of transfer with distance, as predicted by theory [45], this would provide very strong support for the existence of an exchange mechanism. However, this requires that the transfer probability (W) be determined quantitatively for a concrete distance between ions in the lattice, which is a difficult problem. However, the value of such experiments is not limited solely to establishing the existence of a short-range mechanism — they also permit an estimate of its efficiency.

In the multipole interaction studies [42–45], the probability W was calculated using equations in which the average distance between TR^{3+} ion additives were used. These distances were determined starting from the concentration of additive.

The investigations carried out in Chap. 1 enable us to solve the problem of determining W experimentally for specific interionic distances in garnet lattices. Indeed, according to these results it is possible to distinguish in these crystals two types of spectra of the ion additives: a) those having no other ion additives in the first coordination sphere; b) those having at least one ion additive as nearest neighbor.

The possibility of distinguishing the spectra of donors entering into the composition of the activator centers of the type indicated makes it possible to measure individually for each center characteristics such as the relative quantum yield η and lifetime τ, with the aid of which it is then possible to obtain a value for the probability of transfer according to the well-known equation

$$\eta = \frac{(1/\tau_0)}{W + (1/\tau_0)}; \qquad \frac{1}{\tau} = \frac{1}{\tau_0} + W. \tag{10}$$

Here τ_0 is the emission lifetime of the donor.

Since the distance (R) is known for the first and second cationic coordination spheres, it is possible to describe with sufficient accuracy the change in W(R) when going from the first to the second coordination spheres.

Thus knowing the quantity W(R), it is possible both to verify the existence of a short-range energy transfer mechanicm in a given system of TR^{3+} ions and to determine the absolute value of the probability of transfer at the shortest possible distance between the donor and acceptor.

It is clear that the short-range mechanism will have a dominant effect when electron transitions in the donor and acceptor are forbidden, so that the inductive mechanism is ineffective [45].

In this context the question arises of the choice of donor−acceptor pair to be investigated.

We chose crystals of $Y_3Al_5O_{12}$ activated by Eu^{3+} and Yb^{3+} as the object of study, and when the concentration of donors (Eu^{3+}) was not more than 1 at.% the concentration of acceptors (Yb^{3+}) varied from 5 at.% to total replacement of Y^{3+} by Yb^{3+}.

In our opinion, the above $Eu^{3+}-Yb^{3+}$ donor−acceptor pair has a number of advantages in experimental attempts to study the energy transfer mechanism via short-range interaction.

1. The pattern of the energy levels of the Eu^{3+} and Yb^{3+} ions constructed on the basis of an analysis of the luminescence and absorption spectra (Fig. 25) shows that the overlap integral for the emission spectrum of the donor and absorption of the acceptor is nonzero, and the interaction $Eu^{3+} \rightarrow Yb^{3+}$ can occur with emission of a photon with energy $\Delta E > 860$ cm^{-1}.

2. It is seen from Fig. 25 that the transition $^5D_0 \rightarrow {}^7F_6 Eu^{3+}$ will be responsible for the energy transfer, and our measurements indicate that the oscillator strength of this transition is

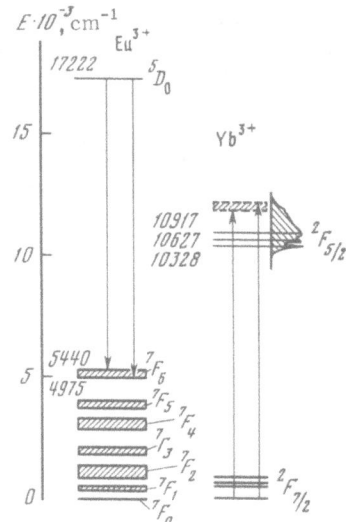

Fig. 25. Splitting of the energy levels of Eu^{3+} and Yb^{3+} in $(Y_{1-x}Yb_x)_3Al_5O_{12}$ crystals at 77°K. Arrows indicate energy resonance possible in the case of energy transfer from Eu^{3+} to Yb^{3+}.

one of the lowest among the radiative transitions $^5D_0 \rightarrow {}^7F_J$ (about 10^3 times less than for the most intense transition $^5D_0 \rightarrow {}^7F_1$). If we take into account that the absolute probabilities of the forbidden transitions $^5D_0 \rightarrow {}^7F_J$ in Eu^{3+} and $^2F_{5/2} \rightarrow {}^2F_{7/2}$ in Yb^{3+} are in turn small (as indicated by the long lifetimes of the metastable levels of these ions, 4.2 and 1 msec, respectively), the relative contribution of the short-range energy transfer mechanism for small distances between the donor and acceptor can be predominant.

3. In the situation considered there is no migration of the excitation energy over the donor ions owing to the extremely small probability of the transition $^5D_0 \rightarrow {}^7F_0$ (according to our data, ≈ 0.1 sec^{-1}) which is responsible for energy transfer among the Eu^{3+} ions at low temperatures. The low probability of the resonance transition $^5D_0 \rightarrow {}^7F_0$ makes the process of transfer over the Eu^{3+} ions practically impossible if the concentration of the latter does not exceed, at the outside, 1 at.%. This fact was established experimentally in crystals of $(Y_{1-x}Yb_x)_3Al_5O_{12} - Eu^{3+}$ in which the concentration of Eu^{3+} was varied from 0.1 to 1 at.%. No reduction of the lifetime of Eu^{3+} was detected.

4. Owing to the special features of the pattern of the energy levels, for the pair $Eu^{3+} \rightarrow Yb^{3+}$ the reverse transfer $Yb^{3+} \rightarrow Eu^{3+}$ with excitation of the level $^5D_0$ of Eu^{3+} is only possible by means of a cooperative process, the probability of which is known [51] to be negligibly small compared to the probability of direct transfer.

Taking into account points 3 and 4, interaction between donors and oscillation effects of excitation energy between the donor and acceptor may be neglected in studying the effects of the transfer $Eu^{3+} \rightarrow Yb^{3+}$.

Since the method for distinguishing AC proposed above is based on studying the dependences of the intensities of the spectral lines of different AC on the concentration of additive, the series of crystals $(Y_{1-x}Yb_x)_3Al_5O_{12} - Eu^{3+}$ in which the concentration x varied from 0.5 to 1 was studied. As control for the positions of the luminescence lines of the Eu^{3+} AC, which due to quenching processes has a very small intensity in the crystals studied, the series $(Y_{1-x}Lu_x)_3Al_5O_{12}$ was used. The Lu^{3+} ions have no absorption bands in the region of transparency of the crystal, and quenching of Eu^{3+} in the $Eu^{3+} + Lu^{3+}$ is absent. At the same time the positions of the lines of the AC $Eu^{3+} + Lu^{3+}$ and Yb^{3+} coincide.

§2. Relationship of the Interaction Processes $Eu^{3+} \rightarrow Yb^{3+}$ to the Structure and Concentration of Associates of Eu^{3+} and Yu^{3+} in Crystals of $(Y_{1-x}Yb_x)_3Al_5O_{12}$

As was expected when we chose our object of study, radiationless energy transfer from Eu^{3+} to Yb^{3+} was experimentally observed in terms of a decrease of the emission intensity and lifetime of the excited Eu^{3+} ions when the concentration of Yb^{3+} increased. Luminescence of Yb^{3+} during excitation into an absorption band of Eu^{3+} was also observed, and was absent in the crystal $(Y_{0.9}Yb_{0.1})_3Al_5O_{12}$ not containing Eu^{3+}.

We consider the relationship between the relative intensity of emission of Eu^{3+} and the structure and concentration of AC present in the crystals with a given concentration of acceptors (Yb^{3+}).

At concentrations of Yb^{3+} up to 30 at.% in crystals of $(Y_{1-x}Yb_x)_3Al_5O_{12}$, lines of two types of AC should for the most part be present in the spectra. These centers are given in the model by an Eu^{3+} ion having no Yb^{3+} neighbor in the first coordination sphere (N-center), and an Eu^{3+} + Yb^{3+} pair (M-center).

As follows from Fig. 26a, which shows the luminescence spectra of the $Y_{0.9}Yb_{0.1}(Lu) \cdot Al_5O_{12}Eu^{3+}$ crystals, emission of both the N- and the M-centers was observed only in crystals containing Lu. In the crystal $(Y_{0.9}Yb_{0.1})_3Al_5O_{12}$, only the "isolated" Eu^{3+} N-centers light up. This indicates that the presence of just one Yb^{3+} ion in the first coordination sphere of Eu^{3+} results in virtually complete quenching. At the same time, the presence of Yb^{3+} in the second and higher coordination spheres of Eu^{3+} has little effect on the luminescence of Eu^{3+}. This was responsible for the successful recording of Eu^{3+} N-lines in $(Y_{1-x}Yb_x)_3Al_5O_{12}$ crystals for Yb^{3+} concentrations as high as 70 at.%, in contrast to the case for crystals of $(Y_{1-x}Lu_x)_3Al_5O_{12} - Eu^{3+}$, whereas the concentration of Lu^{3+} increases so does the intensity of the lines of the Eu^{3+} + Lu^{3+} associates, which already at a concentration of Lu^{3+} of 30 at.% have "squeezed out" the lines of the N-centers. The dependence of the relative intensity of luminescence $\zeta = I_N/I_0$ of the N-centers on the concentration of Yb^{3+} (β) in the crystal is illustrated by curve 2 in Fig. 27.

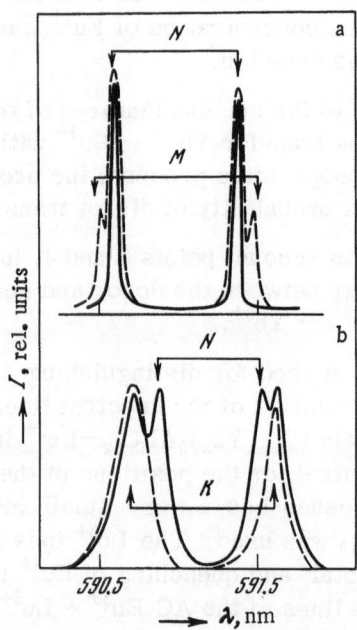

Fig. 26. Luminescence spectra of Eu^{3+} in the group $^5D_0 \rightarrow ^7F_1$ at 4.2°K. a) $Y_{2.7}Lu_{0.3}(Yb_{0.3}) \cdot Al_5O_{12}$; b) $Y_1Lu_2(Yb_2)Al_5O_{12}$ (the scale along the ordinate axis for crystals containing Yb is magnified by $\approx 10^2$ times). The dashed line is for crystals with Lu; the solid line, for crystals with Yb.

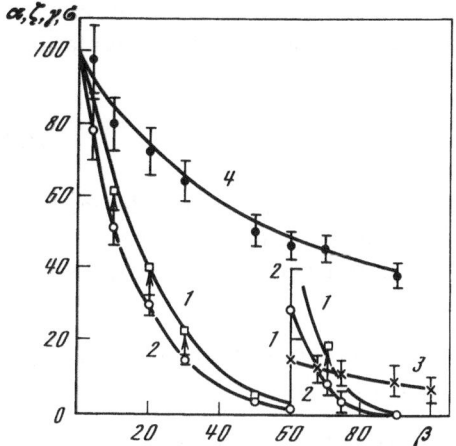

Fig. 27. Dependence of α, ζ, γ, and σ on the total concentration of Yb^{3+}. 1) α, calculated curve for the change in the relative number of N-centers; 2) ζ, change in the relative intensity of luminescence of centers of type N; 3) γ, change in relative intensity of luminescence of K-centers; 4) σ, calculated curve for the change of the relative value of the lifetime of $^5D_0$ of Eu^{3+}; circles, squares, and crosses give experimental points. For $\beta \geq 60$ at.% the scale along the ordinate axis is magnified 20 times. Arrows indicate the change in curve 2 when a correction for quenching of N-centers is introduced.

Here I_N is the integrated intensity of luminescence of the Eu^{3+} N-centers in crystals in the series $(Y_{1-x}Yb_x)_3Al_5O_{12}$; I_0 is the intensity of luminescence of Eu^{3+} in $Y_3Al_5O_{12}$. The concentration of Eu^{3+} in the series of crystals studied was the same. In constructing the function $\zeta(\beta)$ it was possible to ignore corrections for reabsorption, since the terminal level $^7F_1$ of the luminescence transition $^5D_0 \rightarrow {}^7F_1$ studied, which lies 300 cm^{-1} above the ground level, is not populated at 77 and 4.2°K and the oscillator strength of the resonance transition $^5D_0 \rightarrow {}^7F_0$ is vanishingly small.

Figure 27 also shows the dependence of the relative concentration of N-centers on β (curve 1). This curve was calculated using Eq. (9). This equation could be applied because, as shown above, the distribution of TR^{3+} ions in the garnet crystals follows an equiprobable law.

It is seen by comparing the curves $\alpha(\beta)$ and $\zeta(\beta)$ (curves 1 and 2 in Fig. 27) that the change in the relative intensity of luminescence of the N-centers is in good correlation with the change in concentration of N-centers. At the same time, the curve $\zeta(\beta)$ lies somewhat below the curve $\alpha(\beta)$. This indicates that there is some quenching of the Eu^{3+} N-centers caused by interaction with Yb^{3+} located at distances farther away than the Y^{3+} ion, which lies in the first coordination sphere. If this interaction were not present, $\alpha(\beta)$ and $\zeta(\beta)$ would coincide.

The presence of Eu^{3+} AC in the crystals which are characterized by different degrees of quenching should be reflected in experiments measuring the lifetime of the metastable level $^5D_0$ of Eu^{3+}. Thus, the lifetime τ_1 of Eu^{3+} M-centers should be much less than the radiative decay time τ_0, whereas the lifetime of an "isolated" Eu^{3+} N-centers, τ_2, should remain of the same order of magnitude as τ_0.

Curve 4 in Fig. 27 shows the experimentally determined change of $\sigma^{exp} = \tau^{exp}/\tau_0$ with Yb^{3+} concentration. τ^{exp} is the lifetime of Eu^{3+} in $Y_3Al_5O_{12}$. The quantity τ^{exp} was determined from the decay curve of luminescence, which turned out to be exponential over the entire range of variation of Yb^{3+} concentration. As is seen from Fig. 27, the order of magnitude of the change of τ^{exp} remained the same as for τ_0 for a wide range of concentrations of Yb^{3+} up to 90 at.%, i.e., when almost all the Y^{3+} sites in the crystal were occupied by acceptors.

Such a small change permits us to conclude that it is the lifetime of the long-lived N-centers (τ_2^{exp}) that is measured experimentally.

Support for this assertion comes from the fact that if a correction for quenching of N-centers is introduced for the curve $\zeta(\beta)$ in which the value τ^{exp} is used, then the curve $\zeta(\beta)$

coincides with the curve $\alpha(\beta)$ (cf. Fig. 27). The values of the relative quantum yield of luminescence (η_2) of the N-centers for various concentrations were determined for this calculation from the equation

$$\eta_2 = \tau^{\exp}/\tau_0. \tag{11}$$

In studying the kinetics of decay of the metastable level $^5D_0$ in Eu^{3+} we did not detect the presence of quenched M-centers, which can be explained by the fact that owing to the small value of $\tau_1^{\exp}$ for these centers it is not possible to observe them with the equipment we used. According to a preliminary estimate, $\tau_1^{\exp}$ should be ≈ 70 μsec.

Because it was impossible to measure $\tau_1^{\exp}$, the probability of energy transfer between Eu^{3+} and Yb^{3+} (located at minimal distance from Eu^{3+}) was determined by another method.

Luminescence of $Eu^{3+} + Yb^{3+}$ associates could not be detected in the investigation of the spectra of Eu^{3+} in crystals with small Yb^{3+} content because the spectral lines of these associates were of extremely small intensity, both because of the small concentration of such centers, and because their luminescence was strongly quenched. It was not possible to record these lines at the boundary of the intense N-lines. However, as the concentration of the $Eu^{3+} + Yb^{3+}$ associates increased, it became possible to observe the lines of Eu^{3+} centers having at least one Yb^{3+} ion as nearest neighbor (K-centers). This case is illustrated in Fig. 26b for the crystal $Y_1Yb_2Al_5O_{12}$. Here in addition to the N-lines, whose positions are shifted somewhat relative to $Y_3Al_5O_{12}$ due to changes in the unit cell parameters, the lines of the K-centers are also observed. The K-lines were detected in crystals of $(Y_{1-x}Yb_x)_3Al_5O_{12}$ with Yb concentration from 60 to 100 at.%, and the intensity of luminescence of the K-centers is given by curve 3 in Fig. 27. It is seen from the figure that equalization of intensities of the N- and K-lines takes place in a crystal with 67 at.% Yb, where the concentration of Eu^{3+} in the N-centers is $\approx 1\%$ of all the Eu^{3+} ions in the crystal. Naturally, in the case when quenching of Eu^{3+} is eliminated by replacing Yb^{3+} by Lu^{3+} at the same concentration, it was not possible to observe simultaneous luminescence of the N- and K-centers. This case is shown in Fig. 26b, for the crystal $Y_1Lu_2Al_5O_{12} - Eu^{3+}$, in which only K-lines exist in the spectrum.

Using the fact that the intensities of the N- and K-centers are equal in a $Y_3Al_5O_{12}$ crystal with 67 at.% Yb, the value of the relative quantum yield (η_1) of luminescence of centers of K-type can be obtained. It is defined by

$$\eta_1 = \eta_2 (n_2/n_1). \tag{12}$$

Here the ratio of the concentrations of N- and K-centers (n_2/n_1) is equal to $^1/_{99}$ at a concentration of 67 at.% Yb, and the quantity is determined by Eq. (11) with $\beta = 0.67$.

It follows from Eq. (12) that the value of η_1 for Eu^{3+} ions having at least one Yb^{3+} ion in the first coordination sphere is almost two orders of magnitude smaller than the relative quantum yield of Eu^{3+} ions for which there is no Yb^{3+} in the first coordination sphere. Because of this, we can attribute the value of η_1 determined by Eq. (12) exclusively to energy transfer processes from Eu^{3+} to Yb^{3+} located at minimal distance from Eu^{3+}. In this case it is possible to calculate the value of the probability of transfer for an $Eu^{3+} + Yb^{3+}$ pair, if we take into account the various positions of Yb^{3+} in the first coordination sphere of Eu^{3+} for $\beta = 0.67$.

Using the statistical method of calculation, η_1 can be determined by

$$\eta_1 = \left[\sum_{l=1}^{k} \frac{W_0}{W_0 + lW_1} C_k^l \beta^l (1-\beta)^{k-l} \right] : [1 - (1-\beta)^k]. \tag{13}$$

Here k is the coordination number of the first sphere, equal to four for garnet crystals; l is the possible number of Yb^{3+} ions in this sphere (from 1 to 4); C_k^l is the binomial coefficient.

Using Eqs. (11)–(13) with the values of n_1, n_2, η_2 at $\beta = 0.67$, the quantity W_1 turns out to be equal to $\approx (2.5 \pm 1.25) \cdot 10^4$ sec^{-1}. Starting with this value, one can calculate that the lifetime τ_1^{exp} in centers with Yb^{3+} located at minimal distance from Eu^{3+} should be equal to 40 sec. This value is in fact too small to be detected in our experiments investigating the decay kinetics of luminescence.

§3. Short-Range Energy Transfer Mechanism between Eu^{3+} and Yb^{3+} Ions in Crystals with Garnet Structure

As was demonstrated above, the probability of transfer between Eu^{3+} and Yb^{3+} at the shortest distance (R = 3.7 Å) between the ions turns out to be approximately 100 times greater than the probability of radiative decay. It is also possible to estimate the probability of transfer between these ions when the distance between them increases to the radius of the second coordination sphere (5.6 Å). In order to do this the lifetime τ_2^{cal} of Eu^{3+} centers for which the first coordination sphere was not occupied by Yb^{3+} (N-centers) was used.

Two variants must be taken into account in the calculations. First, it is possible that the shortened lifetime in the N-centers with increasing β is due to interaction between Eu^{3+} and Yb^{3+} ions located only in the second coordination sphere. This case is possible if the short-range mechanism of transfer still retains some (small) effectiveness at distances beyond the first coordination sphere. In this case the value of the probability of energy transfer $Eu^{3+} \rightarrow Yb^{3+}$ at the distance of the second coordination sphere, W_2, will have a maximum (W'_2) since in the calculation it is necessary to take into account the average of the transfer to acceptors located only in the second coordination sphere. If quenching of the N-centers is caused by some other mechanism in which there is an interaction involving the acceptors located in some spherical volume around the donors, then in this case the calculated value W''_2 will be smaller than W'_2.

In order to determine the smallest possible value of W''_2 it is necessary to consider the transfer mechanism with the farthest range, the dipole–dipole mechanism. This mechanism is not ruled out for the system considered, since we are concerned here with the so-called stimulated dipole transitions caused by mixing of the electron configurations by the odd crystal field parameters [20].

In this approximation, the transfer probability will on the one hand be proportional to R^{-6}, and on the other hand, its value will remain extremely small. At minimal distances between ions, the efficiency of this mechanism will be low, and quenching can be caused by the short-range mechanism. At larger distances, the existence of a dipole–dipole transfer mechanism must not be ruled out.

The values of W'_2 and W''_2 are calculated in [52] for various positions of the acceptors around the donor, assuming an equiprobable distribution of ions in the d-sublattice sites of garnet. According to these calculations, W' and W'' are equal to 58 ± 20 and 28 ± 10 sec^{-1} respectively. These values are rather large. For this reason, the dependences $\sigma^{cal}(\beta)$ calculated with the aid of the values W'_2 and W''_2 coincide with one another. Consequently, it is not possible to choose one transfer mechanism over the other at the distance of the second coordination sphere by comparing the calculated and the experimental dependences $\sigma^{exp}(\beta)$, $\sigma^{cal}(\beta)$.

The calculated curve $\sigma^{cal}(\beta)$ is shown in Fig. 27 (curve 4), and as is seen from Fig. 27, it is in satisfactory agreement with the experimental results. Thus the decrease of the lifetime of the level $^5D_0$ for Eu^{3+} in N-centers as the concentration of Yb^{3+} increases is explained by the increased number of acceptors in the second and higher coordination spheres around Eu^{3+}.

Using the values of W_1 and W_2 obtained, it is possible to estimate the minimal value of the relative decrease (W_1/W_2) of the probability of transfer $Eu^{3+} \to Yb^{3+}$ when we go from the first to the second coordination spheres. It turns out to be ≈ 200.

In the approximation of an inductive-resonance transfer mechanism for various cases of interaction, the change of W with distance should be proportional to R^{-6}, R^{-8}, or R^{-10}, which correspond to values of W_1/W_2 equal to 13, 30, and 70.

Thus, quantitative estimates of the change of W(R) with distance provide sufficient justification for the assertion that radiationless energy transfer between rare-earth ions such as Eu^{3+} and Yb^{3+} in aluminum garnet crystals cannot occur by one of the multipole mechanisms usually considered.

The efficiency of transfer $Eu^{3+} \to Yb^{3+}$ is virtually restricted to the first coordination sphere, and at the same time, its absolute value remains quite large, $\approx 10^4$-10^5 sec^{-1}, at this distance.

In asserting that in the assumption of a dipole–dipole mechanism the quantity W(R) should vary as R^{-6}, the following fact must be borne in mind which is valid, in particular, for garnet crystals. This fact is that, as follows from [45], the probability of transfer in the dipole–dipole approximation is determined not only by the distance, but also by the quantity $\Phi^2 = 2 \cos \vartheta_1 \cos \vartheta_2 - \sin \vartheta_1 \sin \vartheta_2 \cos \varphi$. Here ϑ_1 and ϑ_2 are the angles between the directions of the dipole moments in the donor and acceptor and the vector joining the two ions; φ is the difference of the azimuthal angles of these dipole moments.

Since these angles will change from sphere to sphere, it is clear that the possibility is not excluded in which Φ^2 = max in the first sphere and is close to zero in the higher coordination spheres. It is just this effect which can cause an abrupt change in W from one coordination sphere to another.

In garnet crystals, such a situation is impossible for the following reasons. First, for rhombic symmetry of the local environment, which is what occurs in garnet (D_2), the dipole moment of transitions in rare-earth ions with half-integral $\mathscr{J}$ cannot have a definite orientation relative to the second-order axes of the nearest-neighbor environment. This conclusion follows from an analysis of the selection rules for the irreducible representations of the group D_2 [53]. The ion Yb^{3+} in our system has a half-integral $\mathscr{J}$. Second, for the Eu^{3+} ion having an integral value of $\mathscr{J}$, orientation of the dipole moment is possible along one of the three second-order axes of its local environment, and the transition $^5D_0 \to {}^7F_6$ corresponds to all three possible directions of the dipole moment. Both these factors cause a certain averaging of Φ^2 even when summation is over a small number of ions located in the acceptor sites nearest the donor.

We calculated the average values Φ_2 over various coordination spheres taking into account the specific features of garnet structure, and it turned out that the ratios Φ_1^2/Φ_2^2 and Φ_1^2/Φ_3^2 are both less than one (the quantities Φ_1^2, Φ_2^2, and Φ_3^2 refer respectively to the first, second, and third coordination spheres), i.e., the quantity Φ^2 in the expression giving the probability of the dipole–dipole mechanism should not decrease, but should rather increase the probability of transfer in going from the first to the second spheres.

Consequently, in a garnet lattice structure the observed abrupt decrease of the probability of energy transfer with distance cannot be caused by "unfavorable" orientation of the dipole moments in the donor and acceptor.

The short-range energy transfer mechanism detected by us experimentally in systems of rare-earth ions can be attributed provisionally to an exchange mechanism or to a mechanism of transfer by means of a virtual exciton. The theory developed in [46, 47, 49] envisages a marked change of the probability of transfer with distance for these mechanisms. As pointed out above, direct overlapping of the wave functions of the TR^{3+} ions must apparently be quite slight. For this reason a "superexchange" mechanism interaction was proposed in [49], in which overlapping of the electron shells of TR^{3+} ions is achieved through intermediate anions. On the basis of the structure of the garnet lattice, it was proposed in [49] that such overlapping can occur in garnets. Apparently, garnet is one of the few structures where superexchange interaction is quite effective. This can explain why in media such as glass no strong quenching of Yb^{3+} by Eu^{3+} ions was detected [54], the mechanism of transfer in this case being attributed to dipole — quadrupole interactions.

As for the transfer mechanism by means of a virtual exciton [46, 47], such transfer is likely to be inefficient because the energy of the excited states of the ion additives in our case differs greatly from the energy of electronic excitation of the crystal.

LITERATURE CITED

1. J. E. Geusic, H. M. Marcos, and L. G. Van Uitert, Appl. Phys. Lett., 4:182 (1964).
2. L. F. Johnson, J. E. Geusic, and L. G. Van Uitert, Appl. Phys. Lett., 7:127 (1965).
3. Kh. S. Bagdasarov, G. A. Bogomolova, D. N. Vylegzhanin, et al., Dokl. Akad. Nauk SSSR, 216:1247 (1974).
4. P. P. Feofilov, Usp. Fiz. Nauk, 58:69 (1956).
5. Yu. K. Voron'ko, V. V. Osiko, A. M. Prokhorov, and I. A. Shcherbakov, Trudy FIAN, 60:3 (1972).
6. A. A. Kaminskii, in: Spectroscopy of Crystals [in Russian], Nauka, Moscow (1973), p. 70.
7. A. A. Kaminskii and V. V. Osiko, Izv. Akad. Nauk SSSR, Neorg. Mater., 1:2049 (1965).
8. W. Low and E. L. Offenbacher, Solid State Phys., 17:135 (1965).
9. F. Euler and J. A. Bruce, Acta Crystallogr., 19:971 (1965).
10. J. A. Koningstein and J. E. Geusic, Phys. Rev., 136A:717 (1964).
11. J. A. Koningstein, J. Chem. Phys., 42:1423 (1965).
12. K. H. Hellwege, S. Hufner, M. Schinkmann, and H. Schmidt, Phys. Kondens. Mater., 4:397 (1966).
13. R. Pappalardo, Z. Phys., 173:374 (1963).
14. M. T. Hutching and W. P. Wolf, J. Chem. Phys., 41:617 (1964).
15. B. F. Dzhurinskii and G. A. Bandurkin, in: Second Seminar on Spectroscopy and Properties of Luminophors Activated by Rare-Earths [in Russian], IRÉ AN SSSR, Moscow (1969), p. 70.
16. W. Low, Paramagnetic Resonance in Solids, Academic Press, New York (1967).
17. K. W. Stivens, Proc. Phys. Soc., A65:209 (1952).
18. G. B. Bokii, L. S. Gaigerova, M. G. Gaiduk, et al., Dokl. Akad. Nauk SSSR, 178:1039 (1968).
19. I. Dabrowski, P. Grunberg, and J. A. Koningstein, J. Chem. Phys., 56:1264 (1972).
20. M. A. El'yashevich, Spectra of the Rare-Earths [in Russian], Gostekhizdat, Moscow (1953).
21. B. Cochane, D. B. Gasson, D. Findlay, et al., J. Phys. Chem. Solids, 29:905 (1968).
22. M. Kestigian and W. W. Holloway, J. Cryst. Growth, 3, 4:455 (1968).
23. R. R. Monchamp, J. Cryst. Growth, 11:310 (1972).
24. R. V. Bakradze, G. M. Zverev, G. Ya. Kolodnyi, and A. M. Onishchenko, in: Spectroscopy of Crystals [in Russian], Nauka, Moscow (1970), p. 222.
25. L. A. Rizeberg and W. C. Holton, J. Appl. Phys., 43:1876 (1972).

26. R. K. Watts and W. C. Holton, J. Appl. Phys., 45:873 (1974).

27. A. Hodges, I. L. Dorman, and H. Makram, Phys. Status Solidi, 35:53 (1969).

28. G. B. Bokii, Crystal Chemistry [in Russian], Nauka, Moscow (1971).

29. A. A. Borovkov, A Course in Probability Theory [in Russian], Nauka, Moscow (1972).

30. G. V. Maksimova, V. V. Osiko, A. A. Sobol', and M. I. Timoshenko, Izv. Akad. Nauk SSSR, Neorg. Materialy, 9:1763 (1973).

31. F. Krieger, Chemistry of Imperfect Crystals [Russian translation], Mir, Moscow (1969).

32. A. M. Prokhorov, V. A. Sychugov, and G. P. Shipulo, Zh. Éksp. Teor. Fiz., 56:1807 (1969).

33. V. A. Sychugov, Author's Abstract of Doctoral Dissertation, FIAN, Moscow (1971). (1971).

34. Yu. K. Voron'ko and V. V. Osiko, Pis'ma Zh. Éksp. Teor. Fiz., 5:357 (1967).

35. K. H. G. Ashbee and G. Thomas, J. Appl. Phys., 39:3778 (1972).

36. M. I. Gaiduk, V. F. Zolin, and L. S. Gaigerova, Luminescence Spectra of Europium [in Russian], Nauka, Moscow (1974).

37. S. Geller, G. P. Espinosa, L. D. Fullmer, and P. B. Crandall, Mater. Res. Bull., 7:1219 (1972).

38. C. D. Brandle, D. C. Miller, and J. W. Nielsen, J. Cryst. Growth, 12:195 (1972).

39. S. Geller, Z. Krystallogr., 12:51 (1967).

40. Yu. K. Voron'ko and A. A. Sobol', Phys. Status Solidi (a), 27:657 (1975).

41. Yu. K. Voron'ko and A. A. Sobol', Fiz. Tverd. Tela, 17:2146 (1975).

42. L. G. Van Uitert, E. F. Dearborn, and J. J. Rubin, J. Chem. Phys., 47:1599 (1967).

43. L. G. Van Uitert, E. F. Dearborn, and J. J. Rubin, J. Chem. Phys., 47:3653 (1967.

44. J. D. Axe and P. I. Weller, J. Chem. Phys., 40:3066 (1964).

45. A. L. Dexter, J. Chem. Phys., 21:836 (1953).

46. V. M. Agronovich, Opt. Spektrosk., 9:798 (1960).

47. M. A. Kozhushner, Zh. Éksp. Teor. Fiz., 56:1940 (1969).

48. L. Rimai, H. Statz, M. J. Weber, et al., Phys. Rev. Lett., 4:125 (1960).

49. L. G. Van Uitert and L. F. Johnson, J. Chem. Phys., 44:3514 (1966).

50. B. M. Antipenko, and V. L. Ermolaev, Opt. Spektrosk., 26:758 (1969).

51. P. P. Feofilov and A. K. Trofimov, Opt. Spektrosk., 27:538 (1969).

52. Yu. K. Voron'ko and A. A. Sobol', Zh. Éksp. Teor. Fiz., 69:1034 (1975).

53. M. Hamermesh, Group Theory and Its Application to Physical Problems, Addison-Wesley (1962).

54. E. Nakasawa and S. Shionoya, J. Chem. Phys., 47:3211 (1967).

INVESTIGATION OF GENERATION OF THE SECOND OPTICAL HARMONIC IN MOLECULAR CRYSTALS*

V. D. Shigorin

Results are presented of a systematic investigation of generation of the second optical harmonic in molecular crystals of organic compounds. The effect of frequency doubling of emission from ruby and neodymium-glass lasers in the powders of 300 organic substances of various classes is studied. Linear and nonlinear optical properties of single crystals of m-dinitrobenzene and m-toluenediamine are investigated. Generation of the second optical harmonic is correlated with the molecular and crystal structures of organic compounds.

INTRODUCTION

Lasers have inaugurated a new stage in the study of the interaction of electromagnetic radiation with matter and have stimulated much work on nonlinear optical effects [1]. The field of nonlinear optics in fact was born in 1961 when Franken et al. [2] experimentally discovered frequency doubling (second harmonic) of ruby laser light in quartz crystals, and it continues to undergo rapid development. The reason for this is that the study of nonlinear optical effects not only leads to a better understanding of the properties of radiation and matter, but also facilitates the development of equipment for regulating laser beams from optically nonlinear materials.

Nonlinear optical properties of a great many crystals have by now been studied, the great majority of the crystals being of inorganic substances. However, considerable interest has recently been expressed in crystals of organic compounds also, which for the most part have a molecular lattice. This is due to the following factors.

1. The extensive possibilities of varying the electron structure of organic molecules interacting weakly in a crystal encourage the hope that a correlation can be established between the structure of the molecules and the properties of the crystals, which would help in explaining the nature of optical nonlinearity and in obtaining the required parameters in crystals.

2. Acentricity of crystal structure, which is necessary for frequency doubling, optical rectification, the linear electrooptical effect, etc. [3], is encountered twice as often in organic as in inorganic substances.

---

*Based on material of a dissertation for the degree of Doctor of Physicomathematical Sciences. The advisors were Academician A. M. Prokhorov and Candidate of Physicomathematical Sciences G. P. Shipulo. The dissertation was defended February 16, 1976 at the P. N. Lebedev Physics Institute of the Academy of Sciences of the USSR.

3. The considerable birefringence of molecular crystals (0.1-0.4) as a rule ensures phase synchronism of interacting light waves, which substantially increases the efficiency of the nonlinear process.

4. The simplicity of synthesizing organic compounds, and the relatively low temperatures used to purify the original material and grow single crystals, makes the solution of a large number of engineering problems easier.

In 1970 we began a systematic investigation of the effect of second-harmonic generation of laser emission (this effect is relatively easy to detect since it is a nonlinear process of low order) in organic molecular crystals, which is described by the quadratic susceptibility tensor

$$\mathscr{P}_i^{2\omega} = \sum_{j,\,k} \chi_{ijk}^{2\omega} E_j^{\omega} E_k^{\omega},$$

where $\mathscr{P}^{2\omega}$ is the polarization of the crystal at twice the frequency of the incoming light wave E^{ω}.

Our investigations were devoted to studying second-harmonic generation in powders and were motivated by a desire to find materials capable of efficiently doubling the frequency of laser light; to study linear and nonlinear optical characteristics of single crystals obtained from some of these substances; and to correlate the nonlinear properties of organic compounds with their molecular and crystal structure. These investigations are described in this paper.

CHAPTER 1

SURVEY OF THE LITERATURE

Molecular crystals are crystals having lattices in which molecules can be isolated as individual structural entities. In such crystals the intermolecular distances exceed by 2-3 times the intramolecular (interatomic) distances, and the lattice energy is usually 1-2 orders of magnitude less than for crystals of any other type of structure. In addition to the universal van der Waals forces, other specific forces acting at short distances, e.g., hydrogen bonds, can act between molecules. However, their directional nature permits the identification in this case of molecules as groups of atoms intimately related to one another. Due to the pronounced dependence of the multipole static interactions on direction, averaging over various orientations of the molecules leads to the result that attraction of the dispersion type and repulsion of electron shells play a principal role even in crystals of polar molecules [4].

Organic compounds form primarily molecular crystals, which are also formed by heteroorganic compounds, binary compounds (CO, HCl), certain simple substances (H_2, N_2, O_2), and elements of the zeroth group. Many crystals to which a molecular structure is usually attributed actually contain, as noted in [5], undirected ionic bonds possessing a specific coordination type.* Such crystals are formed, e.g., by the salts† of organic bases and acids (chlorohydrate, propionate, benzoate, tartrate, oxalate, phthalate, etc.). Thus, crystals of ammonium biphthalate $NH_4HC_8H_4O_4$ are, according to [7], essentially ionic, although phthalic acid $H_2C_8H_4O_4$ itself

---

*The structure of such crystals is molecular-ionic with van der Waals interactions and ionic bonds. It is no longer possible to isolate individual molecules in the lattice.
†The ionic structure of all the salts is pointed out in [6].

forms a molecular lattice [8]. There are no individual structural units in crystals of potassium acid phenylacetate $KHC_8H_6O_2$ either — each cation K^+ is located at the center of a deformed octahedron whose vertices are occupied by the oxygen atoms of the anions of the dimerized acid [9]. On the other hand, in iron pentacarbonyl $Fe(Co)_5$ the metal atom is surrounded by carbonyl groups and the entire molecule is bordered by oxygen atoms, and this compound has a typically molecular lattice [10].

In what follows we consider only organic molecular crystals.

The special features of the structure of molecular crystals show up in their physical properties. Under normal conditions, low melting temperatures ($\lesssim 300°C$), sublimation of whole molecules with heats of sublimation not exceeding a few dozen kcal/mole, large coefficients of compressibility ($\sim 10^{-5}$ atm^{-1}) and thermal expansion ($\sim 10^{-4}$ deg^{-1}), and small density ($\lesssim 2$ $g \cdot cm^{-3}$) and hardness ($\lesssim 100$ $kg \cdot mm^{-2}$) are characteristic of molecular crystals. Most of these crystals are dielectric, although semiconductor properties are observed in numerous organic dyes. In molecular crystals, resonance intermolecular interaction can be observed as a small perturbation of the states of individual molecules, leading to propagation of excitation from molecule to molecule in the form of an exciton (Frenkel exciton). If Davydov splitting is ignored (which does not exceed a few hundred reciprocal centimeters and is observed at low temperatures), the optical spectra of molecular crystals are in general similar to the spectra of free molecules, depend on their electron structure.* and under ordinary conditions have the form of vibronic bands which are broadened and shifted toward longer wavelengths. Moreover, crystals of aromatic or heterocyclic compounds with conjugated bonds have strong absorption ($\sim 10^5$-10^6 cm^{-1}) in the visible or near-ultraviolet region of the spectrum, while crystals of compounds with saturated bonds absorb intensely at $\lambda < 250$ nm. Strong absorption in the near infrared region by organic crystals begins at 1.5-2 μm. According to data of [12], only a few organic crystals are optically isotropic, the great majority having two optical axis in the visible region of the spectrum, with mean index of refraction 1.5-1.7 and birefringence 0.1-0.3. The relatively large value of the birefringence and the acentricity of the unit cell, encountered twice as often in organic than in inorganic substances [13, 14], are important advantages of molecular crystals as potential agents for doubling the frequency of laser light.

It should be emphasized that because of well-known difficulties in obtaining samples of sufficient size and quality, physical properties can only be studied for a small fraction of molecular crystals.

These same difficulties have led in studies seeking out new optically nonlinear crystals to the widespread use of the powder† method, which in many cases permits an estimate of $\chi_{ijk}^{2\omega}$ and establishment of the existence or absence of phase synchronism. The special features of excitation and observation of the second harmonic of laser light in powders will be considered in the following chapter. Here we note that in the powder experiments the average intensities I_2 of the second harmonic detected in the light reflected by or passing through the sample studied, and the second harmonic excited by the same laser beam in a standard sample with known properties, are compared. Such measurements of the relative intensity of second harmonic laser emission in powders were begun in 1964 in [15, 16], which appeared simultaneously with [17, 18] and the first papers in which generation of the second optical harmonic in organic substances was observed. I_2 is correlated with the parameters of a powderlike sample in [19-22].

_____

*The relation between the spectra and the electron structure of molecules is considered, e.g., in [11].

†Crystalline powders were previously successfully used in studying the piezo- and pyroeffect, luminescence, Raman scattering, etc.

The small values of I_2 in the first experiments with powders of methoxy iodide and ethoxy-thiocyanide [15] and a large group of amino acids and sugars [16], or with thin crystal wafers of barbituric acid and benzophenone [15], 1,3-benzanthracene and 3,4-benzpyrene [17], or phenan-threne and benzil [23] during excitation by emission from a ruby laser did not at first attract the attention of workers to crystals of organic compounds as new optically nonlinear materials. However, subsequent measurements of $\chi_{ijk}^{2\omega}$ on single crystals of hexamethylenetetramine (uro-tropin) [18] and hippuric acid [24], in addition to the discovery of substances with efficient gen-eration of the second harmonic of light from a neodymium laser (in the powders of benzyl-monoxime [25], urea and m-dinitrobenzene [20], 7-diethylamino-4-methylcoumarin [26]), indi-cated that the nonlinear optical properties of organic molecular crystals required more exten-sive study.

Just since 1970, second-harmonic generation has been obtained in approximately 200 crystals of compounds of various classes. Radiation of wavelength $\lambda_1 = 1.06$ μm and power dens-ity at the sample ~10^3-10^8 W·cm^{-2} was usually used to excite the harmonic. Here the second harmonic with $\lambda_2 = 0.53$ μm was observed in powders of 3-aminophthalimide [27], m-nitroani-line, numerous derivatives of thiourea [28], benzimidazole [29], certain amino acids and ste-roids [30], 5-nitrouracil [31], 3-bromobenzanthrone, 1-ethoxyanthraquinone [32], disubstituted benzenes [33], p-nitrophenylhydrazine [34], pyrimidine derivatives [35], and numerous other compounds [27-36]. Organic substances most efficient in generating (in powders at T = 300°K) the second harmonic of radiation of a neodymium laser are listed in Table 1. The efficiency of second-harmonic generation in the powders of various compounds with excitation by neodymium or ruby lasers were compared in [38]. In most cases, when the wavelength $\lambda_2 = 0.35$ μm fell in the region of absorption of the crystals, I_2 dropped abruptly, and many materials were destroyed due to two-photon absorption (in the opinion of the authors).

In order to quantitatively determine the tensor components $\chi_{ijk}^{2\omega}$ and other parameters re-lated to second-harmonic generation, it is necessary to have reliable measurements of the in-tensity of the second harmonic, the indices of refraction, and sometimes also the absorption coefficients at λ_1 and λ_2. Such measurements can only be made for sufficiently large oriented single crystal samples of optical quality. Organic single crystals in which second-harmonic generation was studied have been grown in only a few of the papers on this subject, and growth was achieved both from solution (by evaporating the solvent or cooling the solution) and from a melt (method of directed crystallization). Crystallization from solution permits the growth of crystals with fewer stresses at relatively low temperatures, and hence during heating to T_m decomposition of the substances or transition to another polymorphic modification can be avoided. However, it is not always possible to obtain large single crystals from solution, and the much slower rate of growth (as compared to growth from a melt) usually leads to the ap-pearance of impurities in the crystals.

Davydov et al. [39, 40] report the successful growth from a melt (by the Bridgman meth-od) of optically nonlinear single crystals of Anesthesin (15 × 15 × 20 mm), acetamide (15 × 15 × 30 mm), m-nitroaniline (10 × 10 × 40 mm), and m-nitrobenzaldehyde (2 × 2 × 25 mm). Ito et al. [41] recently obtained from a melt single crystals of m-dinitrobenzene of dimensions 20 × 20 × 20 mm. A whole series of organic molecular crystals were grown from solution [39, 40, 42-45] in order to study frequency-doubling of laser radiation with $\lambda_1 = 1.06$ μm. Single crystals of 7-diethylamino-4-methylcoumarin with transverse dimensions as large as 10 mm were obtained in [42] by crystallization from solution. However, due to the presence of impurities from the solvent and other defects, their quality could not be considered satisfactory. Single crystals of m-aminophenol, m-dinitrobenzene, m-dihydroxybenzene, m-nitroaniline, 2-bromo-4-nitroaniline, and 2-chloro-4-nitroaniline were grown in [42] by seeding and isothermal evaporation of the solvent. As indicated by the authors themselves, not one of these crystals was large or of high quality, and many contained small inclusions of solvent or were slightly colored due to impuri-

TABLE 1*

Name	Structural formula	I_2, rel. units	Literature
5-Nitrouracil		4.7 3.3	[31] [35]
Semicarbazone-5-nitro-furfural		4.2	[36]
1-Phenethyl-3-p-nitrophenyl-2-thiourea		3.3	[37]
m-Nitroaniline		1.7 0.7	[40] [31]
4-Nitro-6-methylaniline		1.7	[36]
2-Amino-4-nitro-6-chloro-phenol		1.7	[36]
p-Nitrophenylhydrazine		1.7	[34]
2-Chloro-4-phenyl-5,6-di-hydrobenzoquinazoline		1.3	[35]
7-Diethylamino-4-methyl coumarin		1	[26,27]
3-Bromobenzanthrone		1	[32]
4,5-Diphenyl-3-p-vinyl-pyrazoline		1	[32]
2,4-Dinotrophenol		1	[36]

*The value of I_2 in lithium niobate powder was taken as the unit of measurement of the second-harmonic intensity I_2. The relation pointed out in [36] was used in rescaling I_2 of substances studied in [34-37, 40] to the I_2 of lithium niobate.

ties. As in [42], this situation considerably complicated the method for measuring $\chi^{2\omega}$ and resulted in poorer accuracy. The attempts to obtain crystals from solution in [39, 40, 44, 45] are apparently more successful. In [44], single crystals of 5-nitrouracil of good quality and typical dimensions 2 × 9 × 12 mm were obtained by cooling an aqueous solution from 90 to 45°C at a rate of 1 deg/day in [44]. In [39, 40], crystals of m-nitroaniline (20 × 30 × 45 mm) and picric acid (10 × 30 × 50 mm) respectively were obtained from solutions in mixtures of acetone and

TABLE 2*

Name	Structural formula	Symmetry of crystal	$\chi^{2\omega}_{ijk}$, rel. units	Literature
Hexamethylene-tetramine		$43m$	$\chi^{2\omega}_{xyz} = 10$	[18]
Hippuric acid		222	$\chi^{2\omega}_{zxy} \approx 6$	[24]
Benzil		32	$\chi^{2\omega}_{xxx} = 8.6 \pm 1.2$	[31]
7-Diethylamino-4-methylcoumarin		2	$\chi^{2\omega}_{yxx} = 10$ $\chi^{2\omega}_{yzz} = 1.3$ $\chi^{2\omega}_{xyz} = 1.5$ $\chi^{2\omega}_{yyy} = 3.9$	[42]
5-Nitrouracil		222	$\chi^{2\omega}_{zxy} = 12.4 \pm 2.5$	[44]
d-Threonine		222	$\chi^{2\omega}_{xyz} = 0.96 \pm 0.15$ $\chi^{2\omega}_{yxz} = 1.05 \pm 0.15$ $\chi^{2\omega}_{zxy} = 0.98 \pm 0.15$	[45]
m-Dihydroxybenzene (resorcin)		$mm2$	$\chi^{2\omega}_{zxx} = 1.3 \pm 0.4$ $\chi^{2\omega}_{zyy} = 1.6 \pm 0.5$ $\chi^{2\omega}_{yzy} = 1.2 \pm 0.4$ $\chi^{2\omega}_{zzz} = 1.7 \pm 0.5$	[43]
m-Aminophenol		$mm2$	$\chi^{2\omega}_{zxx} = 3.5 \pm 0.5$ $\chi^{2\omega}_{xzx} = 3.1 \pm 0.5$ $\chi^{2\omega}_{zyy} = 2.5 \pm 0.4$ $\chi^{2\omega}_{yzy} = 2.4 \pm 0.4$ $\chi^{2\omega}_{zzz} = 3.6 \pm 0.5$	[43]
m-Nitroaniline		$mm2$	$\chi^{2\omega}_{zxx} = 51.0 \pm 7.6$ $\chi^{2\omega}_{xzx} = 55.0 \pm 8.2$ $\chi^{2\omega}_{zyy} < 1.1$ $\chi^{2\omega}_{zzz} = 23.0 \pm 3.5$	[43]
2-Chloro-4-nitroaniline		$mm2$	$\chi^{2\omega}_{zxx} = 15.0 \pm 2.2$ $\chi^{2\omega}_{xzx} = 13.0 \pm 2.0$ $\chi^{2\omega}_{zyy} = 35.0 \pm 5.2$ $\chi^{2\omega}_{yzy} = 35.0 \pm 5.2$ $\chi^{2\omega}_{zzz} = 6.5 \pm 1.0$	[43]

TABLE 2 (Continued)

Name	Structural formula	Symmetry of crystal	$\chi_{ijk}^{2\omega}$, rel. units	Literature
2-Bromo-4-nitro-aniline	NH_2 Br NO_2	$mm2$	$\chi_{zxx}^{2\omega} = 15.0 \pm 4.5$ $\chi_{xzx}^{2\omega} = 11.0 \pm 3.3$ $\chi_{zyy}^{2\omega} = 35.0 \pm 10.5$ $\chi_{yzy}^{2\omega} = 28.0 \pm 8.4$ $\chi_{zzz}^{2\omega} = 6.8 \pm 2.0$	[43]
m-Dinitrobenzene	NO_2 NO_2	$mm2$	$\chi_{zxx}^{2\omega} < 2.5$ $\chi_{xzx}^{2\omega} < 2$ $\chi_{zyy}^{2\omega} = 4.5 \pm 2.2$ $\chi_{yzy}^{2\omega} = 3.6 \pm 1.8$ $\chi_{zzz}^{2\omega} = 1.6 \pm 0.8$ $\chi_{zyy}^{2\omega} (\lambda_1 = 1.15 \ \mu m) = (3.1 \pm 0.2) \chi_{36}^{2\omega} (KDP)$	[43] [41]

*The value of $\chi_{zxy}^{2\omega}$ for KDP crystals at $\lambda_1 = 1.06 \ \mu m$ was taken as unit. Its absolute value is $+1.04 \cdot 10^{-9}$ cgse [46, 47]. All the $\chi_{ijk}^{2\omega}$ are rescaled relative to $\chi_{zxy}^{2\omega}$ for KDP using [47, 48]. The Z axis coincides with the principal axis of symmetry of the crystals. The nonlinear susceptibility of crystals of hippuric acid was measured at $\lambda_1 = 0.69 \ \mu m$, that of the remaining crystals, at $\lambda_1 = 1.06 \ \mu m$.

benzene in the ratio $3:1$ and acetone and ether in the ratio $1:1$. Single crystals of p-anisidine ($10 \times 10 \times 40$ mm) were also obtained from solution [39, 40]. Crystals of D-threonine of good optical quality and dimensions $5 \times 10 \times 100$ mm were grown in [45] from a water−ethanol solution. Owing to various defects (dislocations, stresses, etc.) which form during crystallization, molecular crystals usually have a mosaic or block structure, i.e., contain an assembly of sections (blocks) of dimension $\sim 10^{-5}$ cm which have a high degree of structural perfection but are disoriented relative to one another by a fraction of a minute or degree.

The results of the measurements of $\chi_{ijk}^{2\omega}$ for organic molecular crystals (T = 300°K) of which we are aware are given in Table 2.* The values of $\chi_{ijk}^{2\omega}$ for anthracene, acenaphthene, 1,2-benzanthracene, benzil, 1,8-dinitronaphthalene, benzophenone, m-diiodobenzene, m-dinitrobenzene, m-dihydroxybenzene, and phenanthrene are not included, since they were obtained from second-harmonic generation of ruby laser radiation in single crystals of low quality (obtained "over a period of several minutes" from a melt between two quartz disks [49]). It is seen from Table 2 that the $\chi_{ijk}^{2\omega}$ for m-nitroaniline, 2-chloro- and 2-bromo-4-nitroaniline, and 5-nitrouracil are comparable to or greater than the values of $\chi_{ijk}^{2\omega}$ for the ionic crystals of lithium iodate and barium sodium niobate with pronounced optical nonlinearities.

Because of the characteristically large birefringence, many organic substances are capable of giving synchronous frequency doubling to the excitation light. The existence of collin-

*The signs of the $\chi_{ijk}^{2\omega}$ were not determined. In [42, 43] the $\chi_{ijk}^{2\omega}$ were measured on thin single crystal layers, in which the losses were relatively small, owing to the poor quality of the crystals.

ear ($K_1 \| K_2$) phase synchronism during second-harmonic generation with $\lambda_2 = 0.53\,\mu m$ was demonstrated experimentally for single crystals of urea [20], 7-diethylamino-4-methylcoumarin [26, 42], 5-nitrouracil [44], D-threonine [45], m-nitroaniline* [50, 52], Anesthesin [50, 52], acetamide, N,N'-dimethyl-p-nitroaniline, m-nitrobenzaldehyde, and 8-hydroxyquinoline [40], and was predicted using the powder technique for several dozen organic crystals [20, 27, 31, 36]. The possible directions of collinear phase synchronism in optically biaxial crystals (which comprise the overwhelming majority of molecular crystals) were in addition considered theoretically in [53, 54]. Noncollinear (vector) phase synchronism has been much less studied. Experimentally, it has been observed only in single crystals of m-nitroaniline [39, 52, 55], Anesthesin [39, 40, 52], and 8-hydroxyquinoline [39]. General formulas determining all the possible directions of the wave normals of the principal and the transformed radiation inside a biaxial crystal under conditions of vector phase synchronism for generation of the second harmonic have not been obtained.

In addition to significant optical nonlinearities and the possibility of synchronous generation of the second harmonic, a characteristic feature of many organic molecular crystals is their greater stability to laser radiation. Thus, during repeated irradiation by "giant" pulses of a neodymium laser, crystals of 7-diethylamino-4-methylcoumarin withstand power densities of $\sim 100\,MW \cdot cm^{-2}$ [26], and crystals of m-nitroaniline, $200\,MW \cdot cm^{-2}$ [50], whereas crystals of lithium niobate can only take $\lesssim 100\,MW \cdot cm^{-2}$ [56]. Destruction of optical homogeneity of single crystals of 5-nitrouracil [44] and D-threonine [45] is not detected during irradiation by an argon laser with $\lambda = 0.51\,\mu m$ and power density $\sim 1\,kW \cdot cm^{-2}$. Crystals of m-nitroaniline according to data of [50] withstand at least $10\,W \cdot cm^{-2}$ of continuous radiation with $\lambda = 1.06\,\mu m$. The increased stability to radiation by pulses of high power density is also noted in many powders of organic substances [27, 36]. Unfortunately, the softness, ready sublimability, and progressive destruction of surfaces exposed to air characteristic of organic molecular crystals complicates working with them and require, for example, the use during handling of solvents and soft resins, hermetization of prepared samples, etc.

Of fundamental importance is the question of the nature of the optical nonlinearity of organic crystals and its relation to the electron structure of the molecules (which in such crystals retain their individuality to a certain extent). This problem has been considered in papers by various workers. In [23], it is shown, on the basis of an analysis of the π-electron states in conjugated hydrocarbons, that in crystals obtained there (e.g., in phenanthrene), in full accordance with experiment, extremely slight second-harmonic generation is to be expected and could be caused by transitions in which both π- and σ-levels participate. If the conjugation properties are destroyed e.g., by odd-membered rings or by adding substituents, the nonlinear polarizability of the molecule increases, especially for groups of atoms with a strong inductive effect and high refracting power [49]. The use in [57, 58] of variational methods from perturbation theory and a scheme for tensor addition additivity of the polarizabilities of the molecular structural elements made it possible to obtain a satisfactory correlation of the calculated values of $\chi^{2\omega}_{ijk}$ for a hexamethylenetetramine crystal with experiment [18]. It is of interest that the nonlinear polarizability of the unshared electron pair of the nitrogen atom calculated in the LCAO-MO approximation is many times greater than the nonlinear polarizability of the $C-H$ or $C-N$ bond [58], and replacement of the nitrogen atom by a CH group decreases the susceptibility of hexamethylenetetramine by a factor of 2 [57]. According to [44], the optical nonlinearities of

---

*An (energy) conversion coefficient of 15% was obtained in [50] during synchronous generation of second-harmonic radiation from a neodymium laser with $P \sim 200\,MW \cdot cm^{-2}$ in crystals of m-nitroaniline of length 2.5 mm.

molecules of 5-nitrouracil* are caused by the asymmetric distribution of unshared electron pairs of the nitrogen and oxygen atoms.

In [27] the hypothesis was advanced that $\chi^{2\omega}$ increases as the frequency of the second harmonic approaches the frequency of the $S_0 \rightarrow T_1$ transition in the molecules. The double anharmonic oscillator model [43] also gives a dispersion increase of $\chi^{2\omega}$ for a comparatively large effective anharmonicity D of the oscillators of the longest wavelength absorption band, even when the number density ρ of these oscillators is small. However, the use in [43] of identical values of D and ρ in establishing the frequency dependence of nonlinearity in real crystals of compounds characterized by long-wavelength transitions of different natures and with different packing of molecules in the unit cell can hardly be considered to be correct.

Starting from the expression for $\chi^{2\omega}$ for a two-level system and the results of measurement of I_2 in powders of compounds with electron donor or electron acceptor substituents, it was concluded in [28, 30, 37] that large nonlinear susceptibilities (under favorable geometric conditions) are associated with allowed transitions involving charge transfer† which lie nearest to the frequency of the transformed radiation. The contribution of the other transitions is assumed to be much smaller, independently of the position of the absorption bands correspond- to them. The existence of strong electron donor and acceptor groups of atoms in molecules of crystals efficient in second-harmonic generation was also pointed out in [29]. 7-diethylamino-4-methylcoumarin contains such groups, and during excitation by neodymium laser light three-photon absorption in solutions and second-harmonic generation in crystals is recorded in this compound; here a characteristic effect of the polarity of the solvent on the absorption spectra and luminescence of the solution [29] indicated an intramolecular charge transfer in these transitions.

Intense second-harmonic generation in the powders of many organic compounds is probably also related to the above-mentioned structural and spectral features of these molecules. Nevertheless, the consideration of a two-level system with only one transition singled out (transition with charge transfer) cannot serve to explain this relation in the general case. Indeed, the spectra of molecular systems studied in [29, 37] and elsewhere consist of closely spaced bands of differing nature. Therefore, it is possible to limit oneself to only one electron transition at λ_0 in determining $\chi^{2\omega}$ either by introducing a corresponding effective "mean" level‡ corresponding to the transition, or else by assuming the transition is resonance ($\lambda_0 \approx \lambda_2$). However, there is no justification for such assumptions with respect to transitions involving charge transfer: As noted in [37], for many organic substances the mutual position of a charge transfer band with $\lambda_{max} = 0.25$-0.37 μm (in solutions) and $\lambda_2 = 0.53$ μm has no significant effect on the intensity of the second harmonic generated in the corresponding powders.

---

*In 5-nitrouracil the unshared p-electrons of the nitrogen atoms participate in conjugation with the π-electron system of the molecule. This is called p-π-conjugation and the p-electrons in this case are denoted by the letter l. At the same time, the unshared electron pairs of the oxygen atoms do not participate in conjugation with the π-system. Such electrons are denoted by the letter n.

†Every electronic transition m $\rightarrow$ n is accompanied by "charge transfer," i.e., a redistribution of electron density in the system. What we mean here by a transition with charge transfer is an intense transition in which there is a large change in the molecular dipole moment $\mu_{nn} - \mu_{mm}$.

‡Such an effective level (transition) contains, e.g., the one-term Sellmeier formula for the linear susceptibility (or index of refraction) of crystals.

In conclusion,* we emphasize that when our work was started (1970) systematic studies of generation of the second optical harmonic in molecular crystals had not been carried out, and many problems required further study. It was necessary to considerably extend the available experimental material, since $\chi^{2\omega}_{ijk}$ had been measured only for two single crystals (hexamethylenetetramine [18] and hippuric acid [24]), and values $I_2 \sim I_2$ (LiNbO$_3$) had been recorded only for the powder of 7-diethylamino-4-methylcoumarin [26, 27]. It was necessary to establish correlations between the quadratic susceptibility and the molecular and crystal structure, and to do this either quantitatively or semiquantitatively for at least the simplest organic compounds. The problems of noncollinear phase synchronism in biaxial crystals, of estimating the practical possibilities for determining crystal acentricity by observations of second-harmonic generation, etc., were of interest.

CHAPTER 2

INVESTIGATION OF GENERATION OF THE SECOND OPTICAL HARMONIC IN POWDERS OF ORGANIC COMPOUNDS

§1. Second-Harmonic Generation in Crystalline Powders

Due to the difficulties in obtaining sufficiently large single crystals of good quality, it is important to study as a preliminary second-harmonic generation in easily obtained crystalline powders in the search for promising optically nonlinear materials.

Frequency doubling of laser light in powders as statistically inhomogeneous nonlinear media was studied theoretically in [19, 20, 22].

If a plane-parallel slab of thickness L consisting of randomly oriented densely packed grains of powder of nearly identical radius $\hat{r}$ is excited by a laser beam of diameter D ($\hat{r} \ll L \ll$ D), then in the plane wave approximation the nature of the dependence (averaged over the orientations of the powder particles) $\hat{I}_2$ of the second harmonic on $\hat{r}$ inside single crystals of nonscattering grains (immersed in liquid) is essentially different depending on whether the substances do or do not possess phase synchronism. In the first case one can introduce the mean length of coherent interaction $\hat{l}_c \sim \langle |2n_{i2} - n_{j1} - n_{k1}|^{-1}\rangle$ and mean value $\langle(\chi^{2\omega})^2\rangle$ of the square of the nonlinear susceptibility. When $\hat{r} \ll \hat{l}_c$, the electric fields of the second harmonic in the individual grains are correlated and $\hat{I}_2 \sim L\hat{l}_c^{-2}\hat{r}\langle(\chi^{2\omega})^2\rangle$. When $\hat{r} \gg \hat{l}_c$, there is no such correlation, and taking into account the polydispersed nature of the powder, $\hat{I}_2 \sim L\hat{l}_c^2(r)^{-1}\langle(\chi^{2\omega})^2\rangle$. When $\hat{r} \sim \hat{l}_c$, I_2 assumes its maximal value [20]. In the second case, the greatest contribution to $\hat{I}_2$ comes from grains in which the directions of synchronism almost coincide with the direction of propagation of the laser beam. For values of $\hat{r}$ less than the mean effective length of coherent interaction, $\hat{l}^{av}_{c,eff}$, $\hat{I}_2 \sim \hat{r}$, and when $\hat{r} \gg l_{av,eff}$ ($\hat{r} \sim 10^{-2}$ cm), $\hat{I}_2$ has attained its maximal value, is almost independent of $\hat{r}$, and is determined by the corresponding nonlinear interaction coefficients $\hat{d}^2$. The general form of the function $\hat{I}_2(\hat{r})$ in both of the cases described above is shown in Fig. 1. Saturation of the dependence $\hat{I}_2(\hat{r})$ sets in at larger $\hat{r}$ for a single layer of powder from crystals with synchronous frequency doubling, while for crystals without phase synchronism f_2 does not depend on $\hat{r}$ [19].

---

*The general state of the subject of second-harmonic generation in crystals, the fundamentals of molecular spectroscopy, comparative characteristics of ionic and molecular crystals with quadratic susceptibility, and other topics are not discussed here (see, e.g., [59]).

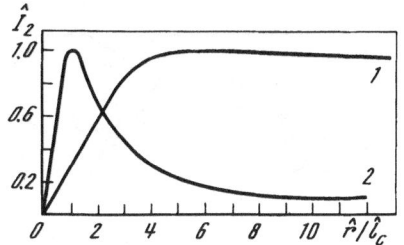

Fig. 1. General form of the dependence of intensity of second harmonic excited in a crystalline powder on the average dimension of the particles in the powder [20]. 1) For crystals with phase synchronism; 2) for crystals without phase synchronism.

When the particles of powder are not immersed in immersion liquid * and are present in air, strong scattering of radiation occurs due to reflection and refraction of light by the grain surfaces, and the angular distribution of the second harmonic is no longer given by a narrow cone along the exciting laser beam, but is described by a cosine (Lambert) law in the "forward" and "backward" directions. However, the character of the function $\hat{I}_2(\hat{r})$ with and without synchronism and the relative values of the total intensity of the second harmonic show practically no change. As $\hat{r}$ increases and the sample becomes more "friable," the reflectivity of a layer of the powder becomes less, and at $\hat{r} \simeq L$ one observes a deviation of the angular distribution of $\hat{I}_2$ from a cosine law which is especially marked for substances with phase synchronism [20].

In molecular crystals of low symmetry, a rather complicated dependence is generally observed of the parameters $\langle(\chi^{2\omega})^2\rangle$ and $\hat{d}^2$ (which are directly related to the intensity of the second harmonic registered experimentally in the powders) on the tensor components $\chi_{ijk}^{2\omega}$. For example, it follows from a general formula for $\langle(\chi^{2\omega})^2\rangle$ in [60] that in weakly dispersive crystals of symmetry mm2 (m-dihydroxybenzene, m-dinitrobenzene, m-nitroaniline)

$$\langle(\chi^{2\omega})^2\rangle = \frac{19}{105}(\chi_{zzz}^{2\omega})^2 + \frac{13}{105}\chi_{zzz}^{2\omega}(\chi_{zxx}^{2\omega} + \chi_{zyy}^{2\omega}) + \frac{44}{105}[(\chi_{zxx}^{2\omega})^2 + (\chi_{zyy}^{2\omega})^2] + \frac{26}{105}\chi_{zxx}^{2\omega}\chi_{zyy}^{2\omega}.$$

A still more complicated expression depending on the symmetry of the crystals (often unknown) and the type of synchronous interactions present is obtained for $\hat{d}^2$.†

On the other hand, when considering the prospects for studying second-harmonic generation in molecular crystals by the powder method, allowance must be made for strong scattering of radiation in organic powders due to the practical difficulties of immersion, for the effect of probable cleavage planes on the grain orientations of a powder with $\hat{r} \geqslant 100\ \mu$m, and for polymorphism.

Thus, in agreement with the experimental data, the relative values of $\hat{I}_2$ make it possible to find only the order of magnitude of $\chi_{ijk}^{2\omega}$ in molecular crystals. However, this, when combined with the possibility of determining in powders the presence or absence of phase synchronism in the crystals and estimating their stability to the effects of laser emission, is entirely sufficient for the purposes of investigations prior to precise measurements on single crystal samples of optical quality.

*Immersion of powders is often made difficult by the lack of suitable highly refractive liquids. For powders or organic compounds, additional difficulties are associated with their solubility. Thus, most of the crystalline substances we studied are soluble in all liquids with n ≥ 1.6 from the standard set of IZh-1 solvents.

†An expression for d in the case of interaction of light waves of type I in crystals of symmetry mm2 is derived in the following chapter.

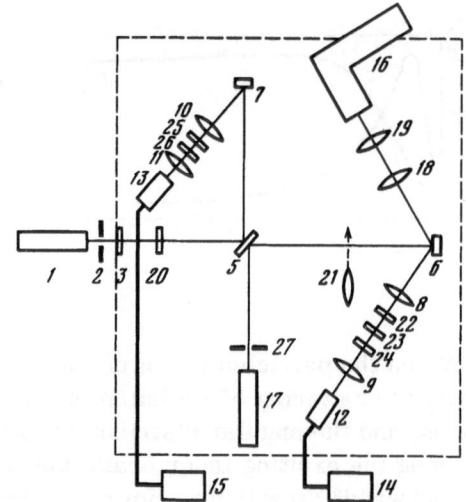

Fig. 2. Diagram of experimental setup for investigating second-harmonic generation laser light in powders. 1) Neodymium (ruby) laser; 2, 27) diaphragms; 3) KS-15 light filter; 4) opaque box; 5) plane-parallel slab; 6, 7) powderlike samples; 8-11, 18, 19, 21) lenses; 12, 13) type FÉU-79 photomultipliers; 14) type S1-19B oscilloscope; 15) type S1-37 oscilloscope; 16) spectral device; 17) type OKG-11 aligning laser; 20, 24) neutral light filters; 22, 25) SZS-21 (UFS-6) light filters; 23, 26) interference light filters at $\lambda = 0.53$ μm (UFS-6).

§2. Experimental Technique for Investigating Second-Harmonic Generation in Powders

Our investigation of second-harmonic generation in powders of organic compounds was carried out using the setup shown in Fig. 2. The neodymium-glass ($\lambda_1 = 1.06$ μm) laser 1 was Q-switched by a rotating prism with total internal reflection (at 150% of the lasing threshold, $P_{max} \simeq 30$ MW, $\tau \simeq 40$ nsec). A single pulse of radiation passed from the laser 1 through the diaphragm 2 selecting the most intense and homogeneous section of the beam, passed through the filter 3, and entered an opaque box 4. Here the laser beam was split by a glas slab 5 into two beams in order to excite the second harmonic in the sample studied 6 and the standard sample 7, which are located at equal distances from the slab. The use of a signal from the standard sample makes it possible to avoid errors in measuring I_2 due to fluctuations of the integrated intensity and mode structure of the main beam [22]. Part of the light flux from the second harmonic from the powderlike samples 6 and 7 was collimated by quartz lenses 8-11 onto photomultipliers 12 and 13, and the second-harmonic signal was recorded on the screens of oscilloscopes 14 and 15. The spectral line of the second-harmonic emission in sanple 6 could be registered at the output of the spectral device 16. A gas laser 17 was used to align the optical elements. The experiments with powders of organic compounds were carried out in several stages.

In the first, preliminary stage, samples were selected in which second-harmonic generation had been observed. Powder from industrial chemical reagents (usually of "pure" or "chemically pure" quality) or from a specially synthesized compound, in which the grain dimensions were 10^{-2}-1 mm, was sprinkled in amounts of a few dozen milligrams into a thin-walled test tube with inner diameter 3 mm. In order to increase the reliability and simplicity in identifying the second harmonic, it was excited by a neodymium laser.* The surface of sample 6 irradiated by this laser was projected by the lenses 18 and 19 onto the slit of an ISP-51 spectrograph. A single green line could be observed visually† or photographically. In order to

---

*The rather bright green light from the second-harmonic radiation from the neodymium laser can often be detected by the unaided eye.

†At a power density at the sample of ~100 MW cm^{-2} in the principal beam and for a spectrograph slit of width 0.3 mm, spectral lines of the second harmonic with $\lambda = 0.53$ μm from the powders for which $I_2 \gtrsim \hat{I}_2$ (SiO$_2$) $\simeq 0.1\hat{I}_2$ (KDP) at $\hat{r} \simeq 100$ μm were recorded visually with confidence.

identify it with the line corresponding to second-harmonic generation of the neodymium laser light, either powder of a known optically nonlinear material (KDP or $LiNbO_3$) or light from an incandescent lamp which had passed through the interference light filter at $\lambda = 0.53$ was used.

In the next stage of the powder investigation, the relative intensity of generation of second-harmonic radiation from the neodymium (or ruby) laser was determined for the substances in which frequency doubling of light with $\lambda = 1.06$ μm had been observed. The industrial reagents were purified by twofold sublimation and the compounds obtained under laboratory conditions were pulverized in a mortar. They were then sifted through sieves with aperture dimensions 75 and 150 μm. The powder so obtained had an average grain* diameter $\hat{r} \simeq 100$ μm and was sprinkled into a flat cell. The cell consisted of two small fused quartz windows held by vaseline against a polished slab of thickness 1 mm with a round excision of diameter 10 mm. The final dense packing of the grains of powder in the cell was achieved before the start of the measurements of I_2 by irradiating the prepared sample with five to six laser pulses (the diameter of the laser beam at the sample was $\simeq$ 6 mm). Dense packing was indicated when the amplitude of the second-harmonic signal from the sample became relatively constant. Copper grids and glasses of type NS calibrated at λ_2 were used to verify that the second-harmonic signal was recorded linearly on the oscilloscope screen. In order to select the second-harmonic radiation, light filters were placed in front of the photomultipliers — in the case of a neodymium laser, an SZS-21 and interference light filters at $\lambda = 0.53$ μm with $T_{max} = 60\%$, $\delta\lambda_{max} = 4$ nm were used; a UFS-6 was used in the case of a ruby laser. The intensity of the second-harmonic generated by the sample under study was compared as a function of the size of the signal with the intensity of the harmonic generated in KDP or $LiNbO_3$ powders with the same $\hat{r}$. The ratio of the amplitudes of the signals from the two samples 6 and 7 did not depend on the power or mode content of the laser emission. The average value of I_2 for this ratio obtained from ten measurements was taken to be the relative value of the intensity of the second-harmonic generated in sample 6. The value of I_2 for sample 6 prepared from the powder of a centrosymmetric crystal (benzoic acid) was essentially zero, while for samples 6 and 7 prepared from a single powder, it was essentially one. When samples were susceptible to destruction by radiation, the laser beam was attenuated by the filters 20.

At the final stage of the investigation of second-harmonic generation, the function $I_2(\hat{r})$ for $\hat{r} \simeq 40, 100, 200,$ and 350 μm was recorded for substances with the largest values of I_2 ($\hat{r} \simeq 100$ μm), and the threshold of destruction of these samples by laser radiation was also studied by gradually increasing the power density of radiation at sample 6 to 400 $MW \cdot cm^{-2}$ using the lens 21. Destruction of the crystals could be judged from the character of the change in the amplitude of the second-harmonic signal with each new laser pulse, and destruction was usually accompanied by breakdown on the surface of the sample and local change of color. If destruction was not observed after repeated irradiation of the sample (~ 100 pulses), the substance was taken to be stable to radiation at the given power density.

All the measurements were made at room temperature.

§3. Determination of the Relative Intensity of the Second Harmonic of Laser Light in Powders of Organic Compounds

We investigated the generation of the second harmonic of laser light in the powders of some 300 organic compounds of different classes [with different types of molecular (σ, π) and atomic (l, n) electron orbitals]. These included the aromatic hydrocarbons (σ, π); aromatic

---

*A microscope was used to monitor the dispersion of the samples.

TABLE 3

Sample No.	Substituents		Position of substituents	I_2, rel. units	Sample No.	Substituents		Position of substituents	I_3, rel. units
	R_1	R_2				R_1	R_2		
1	NO_2	NH_2	meta	42 (1,2)	9	COOH	NH_2	meta	3
2	NO_2	COH	meta	32 (0.9)	10	$COCH_3$	Br	para	2
3	NO_2	NO_2	meta	28 (0.8)	11	NH_2	OH	meta	2
4	NO_2	COH	para	20	12	OH	OH	meta	1.5
5	NO_2	COH	ortho	17	13	NH_2	CH_3	para	0.8
6	NO_2	OH	meta	15	14	NH_2	OCH_3	para	0.7
7	NO_2	CI	meta	5	15	NH_2	OH	para	0.3
8	NO_2	$N(CH_3)_2$	para	4	16	NH_2	NH_2	meta	0.2

amines and phenols (σ, π, l); aromatic nitro compounds, ketones, and quinones (σ, π, n); amino and hydroxy derivatives of aromatic nitro compounds, ketones, and azomethines, oxadiazoles, derivatives of dibenzoylmethane (σ, π, l, n), etc. [61-63]. The effect of frequency doubling was observed in approximately 25% of the substances* studied. For these substances I_2 was measured using excitation by a neodymium or ruby laser.

In the case of neodymium laser the accuracy in measuring I_2 was not worse than 20%, and the divergence in the values of I_2 obtained by us and by other workers [20, 31, 36] for the same substances usually did not exceed 30-40%.†

The relative values of the intensity of second-harmonic generation of a neodymium laser in powders of disubstituted benzenes with the general formula

$$R_1 \diagup\hspace{-0.3em}\bigcirc\hspace{-0.3em}\diagdown R_2$$

are given in Table 3‡ and the values for azomethine compounds with the general formula

$$\begin{matrix} R_2 & R_1 \\ R_3 & \\ R_4 & CH=N-R_5 \end{matrix}$$

are given in Table 4. In addition, the values of I_2 for $\lambda_2 = 0.53$ μm were found for powders of the azo and azomethine compounds shown on the next page,

---

*The crystals of the remaining substances were evidently centrosymmetric, which is in accordance with statistical data [64].

†In particular, according to our measurements I_2 $(SiO_2) = 0.08$ I_2 (KDP), I_2 $(LiNbO_3) = 35$ I_2 (KDP), while at the same time according to [36], I_2 $(SiO_2) = 0.05$ I_2 (KDP), I_2 $(LiNbO_3) = 30$ I_2 (KDP). These data depend on the quality of the standard crystals.

‡All the values of I_2 in Tables 3-5 are expressed in units of I_2 for KDP. The values of I_2 measured directly relative to a $LiNbO_3$ powder with the same $\hat{r} \simeq 100$ μm are given next to these values in parentheses.

TABLE 4*

Sam-ple No.	Substituents					I_2, rel. units
	R_1	R_2	R_3	R_4	R_5	
1	H	H	NO_2	H	⬡–CH_3	315 (9)
2	H	H	H	NO_2	⬡–CH_3	25 (0.7)
3	OH	H	H	H	⬡–$COOC_2H_5$	15
4	OH	H	H	Br	CH_3 ⬡	12
5	H	H	H	H	HN–⬡	12
6	OH	H	H	H	Cl ⬡	10
7	OH	H	H	NO_2	H, C(H_3C)–⬡	8
8	OH	H	H	Cl	H, C(H_3C)–⬡	5
9	H	H	NO_2	H	⬡–$N(CH_3)_2$	4
10	OH	H	H	H	⬡	3
11	H	H	NO_2	H	⬡–$N=CH$–⬡–NO_2	2
12	$NHCOCH_2Cl$	H	H	H	⬡–$COOCH_3$	1
13	OH	NO_2	H	Cl	H_2C–⬡	0.8
14	OH	H	H	Cl	⬡–$N(CH_3)_2$	0.8
15	OH	H	H	H	⬡–$COOC_5H_{11}$	0.1

*Fluorescence in the region 0.54-0.62 μm accompanied by generation of the second harmonic with $\lambda = 0.53$ μm is an interesting property of the powder of compound No. 14 which distinguishes it from all the substances we studied using a neodymium laser.

⬡–NH–N=N–⬡   ($I_2 = 5$)

⬡(NH_2)–N=N–⬡(Cl)   ($I_2 = 0.6$)

⬡–N=N–⬡–$N(CH_3)_2$ ($I_2 = 0.2$)

⬡(OH)–C(CH_3)=N–CH(CH_3)–⬡   ($I_2 = 0.1$)

O_2N⬡–CH=CH–CH=N–⬡–$N(CH_3)_2$ ($I_2 = 0.1$)

as well as for the powders of other compounds of differing classes, which for the most part are listed in Table 5.

When the second harmonic was excited by a ruby laser, an abrupt fall in I_2 (sometimes by several factors of ten) was observed in [62]. Less significant changes in the relative amplitude of the second-harmonic signal are noted only for powders of compounds having no long wave-

TABLE 5

Sample No.	Structural formula	I_2, rel. units
1		245 (7)
2		140 (4)
3		35 (1)
4		28 (0.8)
5		20
6		20
7		15
8		12
9		10
10		8
11		6

TABLE 5 (Continued)

Sample No.	Structural formula	I_2, rel. units
12		5
13		4
14		3
15		3
16		2
17		2
18		2
19		2
20		1.8
21	$H_3C-NH-CH_2-COOH$	1.4
22		0.5
23		0.4
24	$H_2N-CH_2-CH_2-CH_2-CH{\overset{COOH}{\underset{COOH}{}}}$	0.4
25		0.3

TABLE 5 (Continued)

Sample No.	Structural formula	I_2, rel. units
26		0.2
27		0.2
28		0.2
29		0.1
30		0.1
31		<0.1

length absorption bands containing $\lambda_2 \approx 0.35 \ \mu m$. However, in this case also $I_2 \lesssim 10 I_2$ (KDP), in spite of a possible contribution of UV fluorescence transmitted by the UFS-6 light filters with $\delta\lambda_{max} \approx 70$ nm to the recorded signal. Table 6 gives the values of I_2 for some substances excited by neodymium or ruby lasers, as well as the positions of the boundaries λ_b of the long wavelength absorption bands of the crystals $[k(\lambda_b) = 100 \ cm^{-1}]$.

TABLE 6

Structural formula	I_2, rel. units		λ_b, μm
	$\lambda_2 = 0.53 \ \mu m$	$\lambda_2 = 0.35 \ \mu m$	
	315	5	450
	42	6	510 505 *
	32	5	390
	28	4	405 410 *
	20	4	310
	2	3	305 *
	1.5	2	290 *

*An asterisk denotes values of λ_b obtained from [43]. As before, I_2 (KDP) = 1.

TABLE 7

Sample No.	Structural formula	$I_2(\hat{r})$, rel. units			
		$\hat{r} \simeq 40\ \mu m$	$\hat{r} \simeq 100\ \mu m$	$\hat{r} \simeq 200\ \mu m$	$\hat{r} \simeq 350\ \mu m$
1	O_2N—⬡—$CH=N$—⬡—CH_3	4	9	10	10
2		3	7	6	6
3		2	4	4	3.8
4	O_2N—⬡—COH	20	32	32	30
5	H_2N—⬡—CH_3, NH_2	10	20	30	50

These changes in I_2* can apparently be caused by a strong increase in single-photon absorption at λ_2, leading to a smaller I_2 in spite of the dispersive increase of $\chi^{2\omega}$ when λ_2 approaches the resonance value λ_0, when the quadratic susceptibility becomes proportional to the probability of two-photon absorption [65, 66]. Due to the relatively small values of I_2, generation of the second harmonic of ruby laser radiation in the organic powders used by us was not studied further.

§4. Organic Substances of Promise in Obtaining Efficient Frequency Doubling of Laser Light with $\lambda \simeq 1\ \mu m$

The possibility of phase synchronism was established and stability to neodymium laser radiation of a given power density was estimated in order to make a final selection of the most promising of the organic substances studied by us from the viewpoint of obtaining efficient frequency doubling of laser radiation with $\lambda \simeq 1\ \mu m$ in powders with $I_2 \geq 20 I_2$ (KDP).

In accordance with the arguments of Sec. 1 of this chapter, one might expect from the relatively high values of I_2 ($\hat{r} \sim 100\ \mu m$) in the powders of compounds Nos. 1–4 in Table 3, Nos. 1 and 2 in Table 4, and Nos. 1–6 in Table 5, to find phase synchronism for the process of frequency doubling of radiation with $\lambda = 1.06\ \mu m$. The dependence $I_2(\hat{r})$ found confirmed our expectations. The values of the function $I_2(\hat{r})$ for five of the substances investigated are shown in Table 7. For compounds Nos. 1–3 of Table 7, I_2 was measured in terms of the ratio to I_2 (LiNbO$_3$), and for compounds Nos. 4 and 5, in terms of the ratio to I_2 (KDP). In the powder of m-toluenediamine

$$H_2N-⬡-CH_3, \quad NH_2$$

---

*Similar changes in I_2 were also observed in [38] when the neodymium laser was replaced by a ruby laser.

TABLE 8

Name	Structural formula	$I_{2\omega}/I_{2\omega}$ (LiNbO$_3$)	P_{dest}, MW cm^{-2} ($\tau \sim$ 10^{-8} sec)
p-Nitro-p'-methyl-benzalaniline	O_2N-⟨⟩$-CH=N-$⟨⟩$-CH_3$	9	400
p,p-Dibromodi-benzoylmethane		7	300
p-Bromodibenzoyl-methane		4	200
m,m'-Dibromodi-benzoylmethane		1	200
m-Nitrobenzalde-hyde	O_2N⟨⟩COH	0,9	300
1p-Nitrophenyl-5p-methoxyphenyl-3-phenylpyrazoline		0,8	300
m-Dinitrobenzene	O_2N⟨⟩NO_2	0,8	300
m-Nitro-p'-methyl-benzalaniline	NO_2⟨⟩$-CH=N-$⟨⟩CH_3	0,7	400
m-Toluenediamine	H_2N⟨⟩CH_3 NH_2	0,6	300
p-Phenyl-bis-(iminocamphor)		0,6	200

second-harmonic generation of radiation from a neodymium laser is observed which is anomalously intense for a compound possessing only groups of electron donor atoms (cf. Chap. 4). This situation forced us to suspect the existence of a cleavage plane in m-toluenediamine which is almost orthogonal to the direction of synchronism, giving rise to a statistically nonuniform distribution of particles in the powder and "alignment" of the microcrystals along the windows of the cell, so that the incoming laser beam excites significantly more grains in the direction of synchronism than when the cleavage plane is absent. This hypothesis correlated with a frequent steplike increase of the second-harmonic signal when the cell with the powder was placed across a sharply focused laser beam. It also correlated with the observed increase of

$I_2(r)$ up to $\hat{r} \simeq 750\ \mu m.$* The cleavage plane (100) making an angle of 82° with one of the directions of synchronism was indeed found to be present in single crystals of m-toluenediamine that were subsequently grown.† The pronounced cleavage planes may facilitate preferential orientation of the particles in the sample (especially for a single layer of powder and/or large particles) and play an important role in experiments of this type by distorting the results of investigations of second-harmonic generation using the powder method. Cleavage planes of this type are present [12] in, e.g., Anesthesin (100), benzil (001), m-dinitrobenzene (100), urea (110), m-nitroaniline (010), m-nitrophenol (110) and (120), phthalic anhydride (110), and in many other organic crystals for which second-harmonic generation of laser radiation is observed.

In addition to phase synchronism, an important characteristic of optically nonlinear materials is the value of the threshold density of emission P_{dest} above which destruction of the crystal occurs. The determination of P_{dest} for powders was made by the method described in Sec. 2, and because of strong scattering of radiation and defects of the crystal lattice on the surface of the powder grains this determination is only approximate.

Table 8 gives the values of $I_2(\hat{r} \simeq 100\ \mu m)$ and P_{dest} for organic substances found from the results of the investigation to be of promise in obtaining efficient frequency doubling of laser radiation with $\lambda \simeq 1\ \mu m$.

The results of work on the growth of single crystals of some of the organic substances shown in Table 8 are considered in the next chapter, along with a study of second-harmonic generation in them.

§5. Second-Harmonic Generation in Powders as a Method

for Establishing Acentricity of Molecular Crystals

A frequent, albeit auxiliary, problem in the structural analysis of molecular crystals is the important task of determining the acentricity of the unit cell. The very form of the tensor χ^2_{ijk} describing second-harmonic generation suggests that the effect of frequency doubling of laser light might be used in elucidating structures without an inversion center. This possibility was pointed out, e.g., in [20]. In order to assess the practical possibilities of establishing the acentricity of crystals of organic compounds by observing second-harmonic generation of light from a neodymium laser in their powders, we studied [67] second-harmonic generation in 60 organic substances for which the lattice symmetry and the restuls of testing for the piezoelectric effect were known. A neodymium laser and high-aperture ISP-51 ($F_c = 120$ mm) were used in the experiments. The existence of the second harmonic was determined by the quite reliable and simple scheme which we used in the preliminary stage of the investigation of second-harmonic generation in powders and which is described in Sec. 2 of this chapter. The samples were recrystallized from the solvents used in obtaining crystals for x-ray studies‡ and from ethanol.

In most cases there was a good correlation between second-harmonic generation and the symmetry of the crystals.

Acentricity was established by the x-ray method in crystals of m-aminophenol, p-aminophenol, benzophenone, m-dinitrobenzene, 1,8-dinitronaphthalene, m-dihydroxybenzene, xanthene, urea, m-nitroaniline, o-nitrobenzaldehyde, 8-oxyhydroxyquinoline, p-toluidene, m-toluenediamine, and phthalic anhydride (all these compounds have piezoelectric properties [69-71]), as well as diphenylurea, salicylidenaniline, etc. These results were confirmed by the generation in these substances of the second harmonic of neodymium laser light. Among the substances

---

*Analogous effects were not observed for the powders of substances Nos. 1-3 in Table 7.

†Directions of synchronism close to the normal to the cleavage plane were also discovered in crystals of 7-diethylamino-4-methylcoumarin [42].

‡References to the original papers on the crystal structure of substances may be found in [68].

we studied there was not one in which an inversion center was absent (according to x-ray and piezoeffect data), and a line corresponding to the second harmonic with a brightness equal to at least that from crystalline quartz was not observed* (the slit width of the ISP-51 was 0.3 mm). On the other hand, for 4-aminopyridine, xanthone, and p-nitromethylalanine, second-harmonic generation was quite intense (an order of magnitude greater than for quartz) while there wa little piezoelectric effect [70]. This indicates that the method used has a rather high sensitivity.

Many compounds which according to x-ray data crystalline have an inversion center in accordance with this give no generation of the second harmonic)† at $\hat{r} \sim 10$-1000 μm, although most of these compounds possess a polar molecular structure characteristic of second-harmonic generation (cf. Chap. 4). Such compounds include acetanilide, benzoic acid, o-dinitrobenzene, o-dihydroxybenzene, p-nitroaniline, α-nitronaphthalene, p-nitrotoluene, p-nitrophenol, o-nitrochlorobenzene, p-nitrochlorobenzene, and p-phenylenediamine, which have no piezoelectric effect [69], as well as phthalimide, dibenzoylmethane, 4,5-dichloronaphthalic anhydride, naphthalimide, 1-bromonaphthaloxime, and anthraquinonoxy- and anthraquinonethiadiazole.‡

Thus, as is shown by the diffraction method and the method used by us, which give a definitive determination of the symmetry of the average unit cell, the contribution of distorted unit cells on the boundaries of the individual grains of powder and blocks to the total centrosymmetry of the crystals may be neglected.** Second-harmonic generation of light from a neodymium laser was not observed by us either for the above-listed substances when an FÉU-79 and Sl-19B oscillograph were used, which permit reliable recording of second-harmonic signals hundreds of times smaller than from KDP powder.††

However, a number of substances generate a clearly visible green line corresponding to the second harmonic, in contradiction with the centrosymmetry of the crystals of these substances obtained from the same solvents as determined by the x-ray method. Such substances include m-nitrophenol (from benzene), p-aminoazobenzene (from ethanol), p-anisidine (from ethanol), anthra-1,2,5-selenodiazole (from acetic acid), m,m'-dibromodibenzoylmethane (from chloroform), p,p'-dibromodibenzoylmethane (from chloroform), naphthalic anhydride (from chlorobenzene), m-phenylenediamine (from chloroform), and glycine (from water). The intense second-harmonic generation in all these compounds (e.g., for p,p'- and m,m'-dibromodibenzoylmethane, the efficiency of generation was greater than or of the same order as that in $LiNbO_3$ powder with the same grain dimensions) and the piezoeffect in m-nitrophenol [13, 69, 71], p-aminoazobenzene [71], p-anisidine [69, 71], and m-phenylenediamine [13, 69] seemingly indi-

---

*For acentric crystals of conjugated hydrocarbons (phenanthrene, 1,2-benzanthracene, 3,4-benzpyrene, etc.) no second-harmonic generation was observed visually at the power density 100 MW·cm^{-2} of laser radiation used by us. However, second-harmonic generation was reliably detected oscillographically.

†In general, the absence of explicit evidence for second-harmonic generation may be due not only to the centrosymmetry of the crystal, but also to the crystal's belonging to symmetry class 432 or (in the case of weak dispersion) to class 422 or 622, to small values of $\chi^{2\omega}_{ijk}$, to strong absorption at λ_1 and/or λ_2, to low sensitivity of the recording system, etc.

‡The centrosymmetry of the crystals of the last five compounds was established in [72, 73].

**Due to the mosaic structure of the crystals, at the boundary of each block containing n^3 unit cells, $6n^2 - 11n + 8 \approx 6n^2$ ($n \geq 10$) cells are distorted. For a typical block dimension of the order of 100 cells (~ 1000 Å) the relative number of deformed cells on the boundary is a few percent. Impurities in the crystal can increase this fraction.

††Further increase in the sensitivity of the apparatus was contingent upon the possibility of recording second-harmonic signals caused either by local distortions of symmetry near lattice defects, or by magnetic–dipole or electric–quadrupole interactions [74-76].

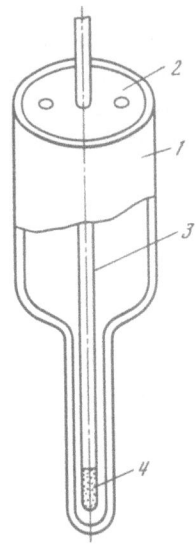

Fig. 3. Dewar for studying the temperature dependence of second-harmonic generation in powders: 1) Dewar; 2) stopper; 3) test tube; 4) powder.

cate either the existence of previously unknown acentric modifications [63] or else an erroneous determination of the space groups of this series of substances as being centrosymmetric [77]. The results of the laser and x-ray method can also differ in the study of crystals which admit motions which reorient molecules or groups of atoms with characteristic times much longer than the duration of the laser pulse [78]. The "anomalous" second-harmonic generation detected by us in crystals of m-nitrophenol and p-anisidine attracted the attention of other workers [79, 80]. The x-ray analysis carried out in [79] for m-nitrophenol showed that when grown from an 80-90% benzene solution, all the single crystals have the symmetry $P2_1/n$, while when grown from 10-20% benzene solution they are assigned to an acentric polymorphic modification with symmetry $P2_12_12_1$. Noncentrosymmetric modifications were also found in the case of p-anisidine [80]. It is thus seen that the study of second-harmonic generation in powders can aid in discovering previously unknown acentric modifications of crystals.

We also studied in [67] second-harmonic generation during phase transitions in thiourea and triglycine sulfate, having Curie points near 173 and 322°K, respectively. The observed temperature dependence of the intensity of the second harmonic clearly reflected the change in symmetry of these crystals with temperature. Although second-harmonic generation could not be observed for thiourea at room temperature, when a test tube containing polycrystals of thiourea was immersed in a Dewar (Fig. 3) containing liquid nitrogen, the green second-harmonic line was distinctly visible spectroscopically, in accordance with the symmetry $P2_1ma$ of the ferroelectric phases. The situation was the opposite when a test tube containing triglycine sulfate was immersed in a Dewar containing hot water — here the crystals acquired the symmetry $P2_1/m$.

Our experiments show that the laser method for determining acentricity of crystals is very simple, permits working with materials in an easily accessible powder form, requires small amounts of sample, and is very quick and sensitive. Observation of second-harmonic generation of laser radiation in crystals is helpful in choosing the most correct of various symmetry groups proposed for them, and it significantly simplifies this problem.* The effect

─────────

*Since second-harmonic generation has by now been studied in the powders of many substances, it seems advisable to take these results into account in the structural analysis of crystals.

of frequency doubling can be used in studying processes accompanied by a change of lattice symmetry involving an inversion center, and the effect on these processes of temperature, pressure, impurities, etc. Second-harmonic generation may find application, e.g., in studies of the temperature dependence of the order parameter during phase transitions of the "order–disorder" type characteristic of organic crystals, and also in monitoring the process of formation of crystalline domains in polymers with a polar backbone (cellulose, polyamides, complex polyethers, etc.).

CHAPTER 3

GENERATION OF THE SECOND-HARMONIC OF LASER LIGHT IN SINGLE CRYSTALS OF m-DINITROBENZENE AND m-TOLUENEDIAMINE

§1. Growth of Single Crystals and Preparation of Samples for Investigation

After substances with efficient frequency doubling of laser radiation were established, it became of great interest to obtain single crystals of the substances of the required dimensions and quality in order to further study quantitatively the generation of the second optical harmonic in these substances.

Attempts to grow such crystals were made for a large number of compounds. However, single crystals of p- and m-nitro-p'-methylbenzalaniline, p-, p'- and m-, m'-dibromodibenzoylmethane, p-bromodibenzoylmethane, m-nitroaniline, m-nitrobenzaldehyde, m-dinitrobenzene, and m-toluenediamine grown from solutions* unfortunately did not attain the required dimensions and/or quality. When single crystals were grown from a melt (by the Bridgman method), even after the test tube containing the substance was evacuated to 10^{-3} mm Hg or filled with argon, decomposition of p- and m-nitro-p'-methylbenzalaniline, p,p'- and m,m'-dibromodibenzoylmethane, and p-bromodibenzoylmethane occurred. At the same time, rather large single crystals were grown in the case of m-nitroaniline, m-nitrobenzaldehyde, m-dinitrobenzene [63], and m-toluenediamine.

Since the single crystals of m-nitroaniline obtained by us coincided with those used in [51, 55] to study the nonlinear optical properties of these crystals,† while the crystals of m-nitrobenzaldehyde contained a large number of defects, we chose for our study of second-harmonic generation crystals of m-dinitrobenzene‡ and m-toluenediamine [89, 90]. It should

---

*The crystals were grown from solutions and melts at the Institute of Crystallography and Institute of Solid State Physics of the Academy of Sciences of the USSR.

†Subsequent papers on single crystals of m-nitroaniline were devoted to observation in them of the electrooptical effect [81-85] and second-harmonic generation of laser light with $\lambda_1 = 1.06\ \mu$m [43, 50, 52].

‡Soon after the publication of our results for these crystals [86], the paper [43] appeared in which optical nonlinearity of a number of benzene derivatives including crystals of m-dinitrobenzene was studied. However, as the authors themselves point out [43], these crystals were of poor quality, which inevitably affected the accuracy of the measurements. Phase synchronism in crystals of m-dinitrobenzene under conditions of second-harmonic generation of laser radiation with $\lambda_1 = 1.15\ \mu$m was recently considered in [41]. A comparison of the results obtained by us [87, 88] and [41, 43] will be given in the following sections of this chapter.

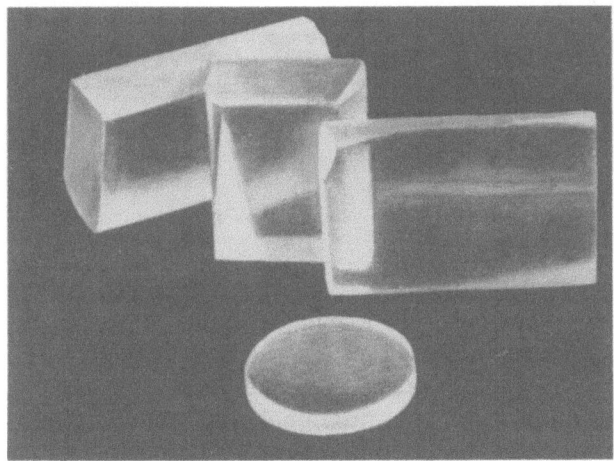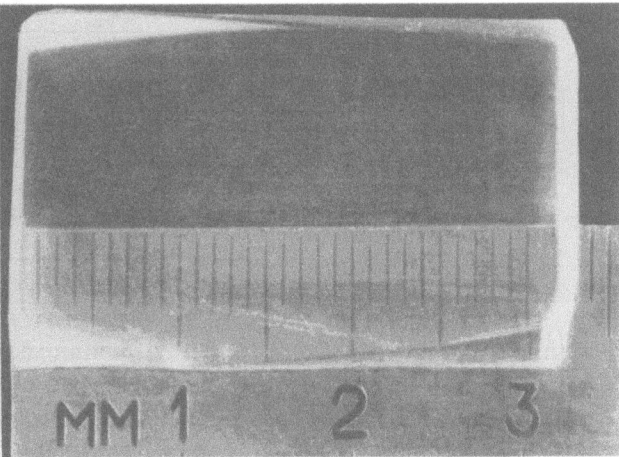

Fig. 4. Single crystals of m-dinitrobenzene and m-toluenediamine.

be noted that crystals of m-dinitrobenzene were previously grown only from solution and their dimensions did not exceed a few cubic millimeters [91-93]; information concerning growth of crystals of m-toluenediamine is absent from the literature. For this reason, special methods for growing large optically homogeneous single crystals of m-dinitrobenzene and m-toluene-diamine were developed [94, 95] which made it possible to obtain after 2-3 weeks crystals of length 30-50 and diameter 15-20 mm. The external appearance of the crystals of m-dinitro-benzene and m-toluenediamine is shown in Fig. 4, and some of their parameters are given in Table 9.

The grown crystals were cut on a Parvov saw (Special Instruments Division, Institute of Crystallography of the Academy of Sciences) using a wire wetted in dimethylformamide or toluene. Using the orthoscopic charts of [98], the portions of the crystals with the fewest optical inhomogeneities caused by internal stresses, intergrowths, bubbles, cracks, etc., were selected in thin (~ 1 mm) slabs in order to prepare the samples. The dislocation density was $\sim 10^6$ cm^{-2} for single crystals of m-dinitrobenzene used in preparing the samples, and the content of impurities was $\lesssim 10^{-2}$ cm^{-2}. The corresponding values for m-toluenediamine were $\sim 10^5$ and $\lesssim 10^{-3}$ cm^{-2}. The microhardness in the planes (001), (010), and (100) was respectively 10, 7, and 8 kg$\cdot$mm^{-2} for crystals of m-dinitrobenzene and 25, 30, and 28 kg$\cdot$mm^{-2} for crystals of m-toluenediamine.* Splitting of the crystals was especially easy in the cleavage plane (100).

The anisotropy of the mechanical and optical properties of the crystals was used in orienting them in addition to x-ray (goniometer) techniques. Characteristic cracks along the (100) cleavage plane were formed when the crystals were rapidly cooled by wetting with a readily evaporating liquid (ether, dichloroethane) or subjected to pressure from a needle directed perpendicular to the cut surface, and the "tails" of the pricking patterns in certain sections indicated corresponding crystallographic directions [e.g., the 010 axis in the (100) plane for crystals of m-dinitrobenzene]. Conoscopic figures observed using an MIN-8 polarizing microscope made it possible to fix the location of the planes (N_g, N_m), and (N_p, N_g) of the optical indicatrix more accurately, which correspond to the planes (001) and (010) of m-dinitro-benzene crystals and (100), (001) of m-toluenediamine crystals.

----

*The microhardness was measured using an MMT-3 device with load 50 g at temperature 25°C. For comparison, the microhardness of KDP crystals is ~ 100 kg mm^{-2} [99].

TABLE 9

Crystal	Space group symmetry	a, Å	b, Å	c, Å	z	ρ, g·cm^{-3}	T_m, °C	Literature
m-Dinitro-benzene	$Pbn\,2_1$	13.257	14.048	3.806	4	1.57	91	[91, 93, 96]
m-Toluene-diamine	$Pna\,2_1$	8.359	12.406	6.225	4	1.26	98	[97]

In preparing the samples for measurement the crystals were ground on glass covered with silk using the solvent (dimethylformamide), and were polished on a soft (No. 12) resin using chromium oxide as the abrasive. The purity of the surfaces of the crystals was of class 9 for m-dinitrobenzene and class 12 for m-toluenediamine. The optical quality of the samples was monitored by radiation from a He−Ne laser. In order to avoid progressive darkening of the polished surfaces of the crystals observed in air over a 24-hour period, the samples were kept in an airtight container.

§2. Dispersion of the Indices of Absorption and Refraction

of m-Dinitrobenzene and m-Toluenediamine Crystals in

the Visible and Near Infrared Regions of the Spectrum

The parameters determining generation of the second optical harmonic are closely related to the dispersion of the linear indices of absorption and refraction of the medium.

Crystals of m-dinitrobenzene and m-toluenediamine belong to the rhombic crystal system and are biaxial with three principal indices of absorption k_i and refraction N_i corresponding to polarization of light along the principal axes i (i = x, y, z) of the crystal.*

In order to reduce the effect of the quality of treatment of the surfaces of the crystals on the results of the measurements, k_i were found from the transmission $T_i(\lambda)$ in the polarization spectra of samples of different thickness L using the formula

$$k(\lambda) = (L_2 - L_1)^{-1} \ln \frac{T_1(\lambda)}{T_2(\lambda)}.$$

Oriented slabs of the crystals studied of thickness 4-10 mm and an SF-8 spectrometer with polarizing prisms were used.

Figure 5 shows the spectra of the indices of absorption of crystals of m-dinitrobenzene and m-toluenediamine.

The principal indices of refraction were determined for each of the crystals by the method of [100, 101] using two triangular prisms, one of the boundaries of which (of dimension 10 × 15 mm) coincided with the symmetry planes ZX and ZY of the optical indicatrix. The indices of refraction were calculated using trigonometric tables [102] by the formula

$$N = \frac{\sin(\alpha + \varphi)}{\sin \alpha},$$

where α is the refracting angle of the prism and φ is the angle of deflection of a light ray by the prism; α and φ were measured at room temperature on a GS-5 goniometer for 10 spectral

*For the rhombic system the directions of the physical and crystallographic axes in the coordinate systems coincide; with the standard choice of axes, 2 ∥ Z ∥ c.

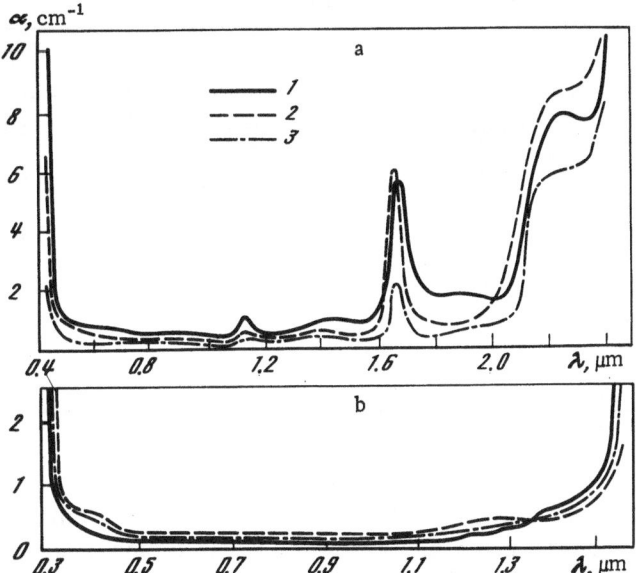

Fig. 5. Dispersion of the indices of absorption k (in cm^{-1}) for crystals of m-dinitrobenzene (a) and m-toluenediamine (b) for radiation polarized: 1) along the X axis; 2) Y axis; 3) Z axis.

radiation lines from mercury and sodium lamps, from a helium−neon laser, and also from an aluminum garnet laser containing neodymium and its second harmonic. Taking into account the errors in measuring the angles ($\leqslant 10"$) and inaccuracies in cutting the prisms ($\leqslant 1°$), the error in determining the indices of refraction [103] did not exceed $\pm 5 \cdot 10^{-4}$. For both crystals the values of N_z measured independently using two prisms were the same, to within the above error.

The indices of refraction of crystals of m-dinitrobenzene and m-toluenediamine are shown in Tables 10 and 11, respectively. As is seen from Table 10, the values of N_i obtained by us for crystals of m-dinitrobenzene differ considerably from the values of other workers.

TABLE 10*

λ, μm	N_x	N_y	N_z
0.436	1.8025	1.7361	1.5072
0.492	1.7731	1.7104	1.4964
0.532	1.7592	1.6983	1.4912
0.546	1.7553	1.6950	1.4896
0.576	1.7483	1.6889	1.4871
	1.7516[41]	1.6853[41]	1.4905[41]
0.577	1.7480	1.6886	1.4869
0.579	1.7476	1.6882	1.4865
0.589	1.7456	1.6865	1.4859
	1.71 [91]	1.68 [91]	1.48 [91]
	1.839 [104]	1.766 [104]	1.432 [104]
	1.841 [105]	1.746 [105]	1.482 [105]
0.633	1.7381	1.6798	1.4827
0.872	1.7152	1.6602	1.4732
1.064	1.7093	1.6539	1.4707
1.153	1.7072	1.6520	1.4698
	1.7068[41]	1.6477[41]	1.4719[41]

*The values of N_i at $\lambda = 0.576$ and 0.872 μm are obtained by interpolation.

TABLE 11*

λ, μm	N_x	N_y	N_z
0.436	1.6433	1.8632	1.8019
0.492	1.6296	1.8320	1.7778
0.532	1.6226	1.8189	1.7676
0.546	1.6205	1.8150	1.7644
0.576	1.6164	1.8072	1.7585
0.577	1.6163	1.8071	1.7583
0.579	1.6161	1.8069	1.7579
0.589	1.6150	1.8047	1.7564
0.633	1.6108	1.7967	1.7499
0.872	1.5975	1.7730	1.7308
1.064	1.5930	1.7644	1.7240
1.153	1.5916	1.7618	1.7220

*The values of N_i at $\lambda = 0.576$ and 0.872 μm are obtained by interpolation.

In [43], unfortunately, the values of N_i for these crystals are not tabulated. However, the indices of refraction found for the interval $\lambda = 0.5$-1.1 μm in terms of the dispersion curves in |43| are almost equal to our values only for N_x and exceed them on the average by $2.5 \cdot 10^{-2}$ for N_y and $2 \cdot 10^{-3}$ for N_z. It follows from the data of |41| that the indices of refraction obtained there for m-dinitrobenzene at $\lambda = 0.45$-1.15 μm on the average exceed our values by $1.5 \cdot 10^{-3}$ for N_x and $3.5 \cdot 10^{-3}$ for N_z, but are smaller by $4 \cdot 10^{-3}$ for N_y; in addition, the largest spread of the experimental points is observed for the dispersion curve $N_y(\lambda)$.

Since the maximal (N_g), average (N_m), and minimal (N_p) indices of refraction of crystals of m-dinitrobenzene and m-toluenediamine satisfy the relation $N_g - N_m < N_m - N_p$ for $\lambda \simeq 0.4$-1.2 μm, in this interval of wavelengths the direction N_p of the optical indicatrix bisects the acute angle

$$2V' = 2 \operatorname{arctg} \left(\frac{N_m^{-2} - N_g^{-2}}{N_p^{-2} - N_m^{-2}} \right)^{1/2}$$

between the optical axes lying in the plane of (N_p, N_g).*

§3. Nonlinear Susceptibility of Crystals of m-Dinitrobenzene and m-Toluenediamine for Frequency Doubling of Laser Light with $\lambda = 1.064$ μm

The vector components of the quadratic polarization $\mathscr{P}_i$ of crystals of symmetry point group mm2, to which crystals of m-dinitrobenzene and m-toluenediamine belong, are related to the components of the electric field vector E_i of the main radiation by

$$\mathscr{P}_x = 2\chi_{xxz}^{2\omega} E_x E_z,$$
$$\mathscr{P}_y = 2\chi_{yyz}^{2\omega} E_y E_z,$$
$$\mathscr{P}_z = \chi_{zxx}^{2\omega} E_x^2 + \chi_{zyy}^{2\omega} E_y^2 + \chi_{zzz}^{2\omega} E_z^2.$$

---

*The indices of refraction N_g, N_m, N_p correspond to oscillations of the electric field vector of a light wave along the axes, X, Y, and Z, respectively, for crystals of m-dinitrobenzene, and along the Y, Z, and X axes for crystals of m-toluenediamine.

There are at present several methods for determining the tensor components $\chi^{2\omega}$ of the nonlinear susceptibility tensor [106-110]. However, in the case of the crystals studied by us, it is in principle possible to find only $\chi^{2\omega}_{zzz}$ by measuring the oscillations of I_2 during interference of the free and stimulated waves of the second harmonic in plane-parallel slabs (the so-called Maker oscillations [106]), while the Maker method [108] modified for optically biaxial crystals is not sufficiently accurate and involves the calculation of auxiliary parameters. On the other hand, observation of the second harmonic in the directions of phase synchronism [109] makes it possible, assuming the fulfillment of the additional symmetry conditions of Kleinman [111], to determine only the two components $\chi^{2\omega}_{zyy} \approx \chi^{2\omega}_{yyz}$ for crystals of m-dinitrobenzene and four components $\chi^{2\omega}_{zxx} \approx \chi^{2\omega}_{xxz}$ and $\chi^{2\omega}_{zyy} \approx \chi^{2\omega}_{yyz}$ for crystals of m-toluenediamine.* On the other hand, the wedge method [110] is simpler and of more general applicability, and does not require high accuracy in preparing the samples and orienting them relative to the laser beam. This method has found wide application in nonlinear optics and was used in our experiment.

Since serious experimental difficulties and large inaccuracies are involved in finding the absolute values of the $\chi^{2\omega}_{ijk}$, we performed relative measurements of the $\chi^{2\omega}_{ijk}$ by comparing the intensities of the second harmonic I_2, their oscillation factors C, and the coherent lengths and indices of refraction N for wedges with angles γ at the vertex made from the crystals studied and from standard KDP crystals. Under the assumption that the power, cross section, and mode structure of the laser beam on the wedgelike samples were constant during the measurement process, and taking into account also the insignificant role of multiple reflections and the smallness of the absorption at frequencies and 2ω ($k_1 L$, $k_2 L \ll 1$), the quantities $\chi^{2\omega}_{ijk}$ of interest to us are equal to

$$\frac{|\chi^{2\omega}_{ijk}|}{\chi^{2\omega}_{zxy(KDP)}} = \left(\frac{I_2}{I_2^{KDP}}\right)^{1/2} \left(\frac{C^{KDP}_{zxy}}{C_{ijk}}\right)^{1/2} \frac{l^{KDP}_{zxy}(N_{i2}+1)(N_{j1}+1)(N_{k1}+1)}{l_{ijk}(N^{KDP}_{e2}+1)(N^{KDP}_{O1}+1)^2},$$

where $\chi^{2\omega}_{zxy}$ (KDP) = $+1.04 \cdot 10^{-9}$ cgse [46, 47]. For the wavelength $\lambda = 1.064$ μm at which the measurements were made, the frequency of the main radiation and its second harmonic lie in the region of transparency of the crystals. Therefore, all the $\chi^{2\omega}_{ijk}$ have real values and fulfillment of the additional symmetry conditions of Kleinman [111] is to be expected for them.

Bearing in mind that for $\lambda_1 = 1.064$ μm $N^{KDP}_{O1} = 1.49375$, $N^{KDP}_{e2} = 1.47045$, and $l^{KDP}_{zxy} = 11.43 \pm 0.02$ μm [107], we have

$$\frac{|\chi^{2\omega}_{ijk}|}{\chi^{2\omega}_{zxy(KDP)}} = 0.74 \left(\frac{I_2}{I_2^{KDP}}\right)^{1/2} \left(\frac{C^{KDP}_{zxy}}{C_{ijk}}\right)^{1/2} \frac{(N_{i2}+1)(N_{j1}+1)(N_{k1}+1)}{l_{ijk}}.$$

It is seen from this expression that the errors in determining the relative values $|\chi^{2\omega}_{ijk}|$ are caused mainly by the overall error in measuring the amplitude of the second-harmonic signal, which is due both to inaccuracy in recording I_2 and to fluctuations of I_2; the departure of I_2/I_2^{max} from the constant value 0.5 attained when the interference oscillations of I_2 are averaged completely over the cross section of a (Gaussian) laser beam [110]; and to errors incurred in finding the coherent lengths.

We used wedges with angles $\gamma = 1$-$2°$. The diameter of the laser beam on the sample was approximately 2 mm. For greater accuracy, the coherent lengths l_{ijk} were calculated from the

---

*The results of this method as applied to crystals of m-dinitrobenzene and m-toluenediamine are given in Sec. 5 of this chapter.

TABLE 12

ijk	m-Dinitro-benzene	m-Toluenediamine
xxz	1.56 ± 0.01	7.42 ± 0.20
yyz	1.96 ± 0.01	3.56 ± 0.04
zxx	1.22 ± 0.01	1.52 ± 0.01
zyy	1.64 ± 0.01	$83 \pm 26;\quad 82 \pm 1$ (expt.)
zzz	12.96 ± 0.65	6.08 ± 0.14

indices of refraction by

$$l_{ijk} = 0.5\lambda_1 \,|\, 2N_{i2} - N_{j1} - N_{k1} \,|^{-1}.$$

The values of l_{ijk} are given in Table 12.

Under the conditions of our experiments, the errors δ_{ijk} in determining the relative values $|\chi_{ijk}^{2\omega}|$ did not exceed 20% for all the tensor components of both crystals studied, except for $\chi_{zzz}^{2\omega}$ for m-dinitrobenzene ($\delta_{zzz} \simeq 25\%$) and $\chi_{zyy}^{2\omega}$ for m-toluenediamine ($\delta_{zyy} \simeq 75\%$). The error in finding the latter component was increased four times when allowance was made for the pronounced oscillations on I_2 in this case when the refracting edge of the wedge was moved perpendicular to the laser beam and the amplitude and period of the oscillations were measured:

$$\varkappa = 2l_{zyy} \tan^{-1} \gamma.$$

The orientation of the wedges of m-dinitrobenzene, m-toluenediamine, and KDP relative to the crystallographic coordinate axes is shown in Fig. 6, and the direction of propagation s_1 of the laser beam, the polarization e_1 of the main radiation, and that of its second harmonic, e_2, during measurement of $\chi_{ijk}^{2\omega}$ in crystals of m-dinitrobenzene and m-toluenediamine are given in Table 13.

The wedges were cut both from individual single crystals and from prisms of the crystals used previously to find the indices of refraction. Typical dimensions of the polished working surfaces were 20×15 mm. All the measurements were performed on the experimental setup shown in Fig. 7. After I_2 was recorded from the wedges of m-dinitrobenzene and m-toluene-diamine, a KS-15 light filter was placed behind the wedges in the path of the laser beam which absorbed the second-harmonic radiation emitted from the wedges but transmitted the main beam, together with a wedge of KDP crystal. This method aided in eliminating as far as possi-

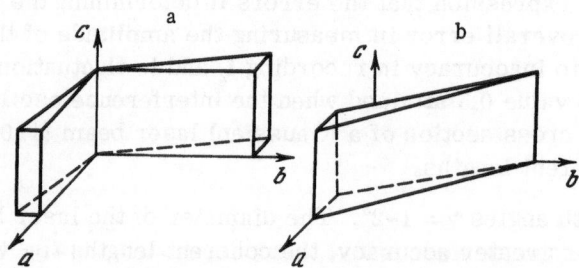

Fig. 6. Orientation of the wedges during measurement of $\chi_{ijk}^{2\omega}$: a) for crystals of m-dinitro-benzene and m-toluenediamine; b) for crystals of KDP.

TABLE 13

ijk	S_1	e_1	e_2
Crystals of m-dinitrobenzene and m-toluenediamine			
xxz	[010]	[101]	[100]
yyz	[100]	[011]	[010]
zxx	[010]	[100]	[001]
zyy	[100]	[010]	[001]
zzz	[100]	[001]	[001]
KDP crystals			
zxy	[110]	[$\overline{110}$]	[001]

ble the effect of the quality of treatment of the organic crystals on the measurements of the relative values of I_2. At power densities of the laser beam at the samples of ~ 1 MW cm^{-2}, the signal-to-noise ratio was at least 10. The laser power was monitored in terms of the intensity of the second harmonic generated in the KDP wedge before and after I_2 was determined for the crystal studied. The system was aligned using image converters and the second-harmonic light ray generated with $\lambda = 0.53$ μm. The nondependence of I_2 (within the accuracy of measurement) on the position of the laser beam along the refracting edge of a given wedge or of different wedges made from the same material for identical crystallographic orientations and values of the angle γ indicated that the samples were optically homogeneous.

The relative values $|\chi^{2\omega}_{ijk}|$ and the Miller coefficients [112] $|\Delta^{2\omega}_{ijk}| = (4\pi)^3 |\chi^{2\omega}_{ijk}|$ $\{(N^2_{i2}-1)(N^2_{j1}-1)(N^2_{k1}-1)\}^{-1}$ found by us for crystals of m-dinitrobenzene and m-toluenediamine at $\lambda_1 = 1.064$ μm are collected in Table 14. The values of these parameters obtained independently in [43] for crystals of m-dinitrobenzene are also given. The discrepancies do not exceed the error of measurement in [43]. The value of $|\chi^{2\omega}_{zyy}|$ recently determined in [41] for m-dinitrobenzene at $\lambda_1 = 1.153$ μm (cf. Table 2) essentially coincides with ours if we bear in mind the negligibly small dispersion of $\chi^{2\omega}$ in the range $\lambda_1 = 1.06$-1.15 μm and the relation

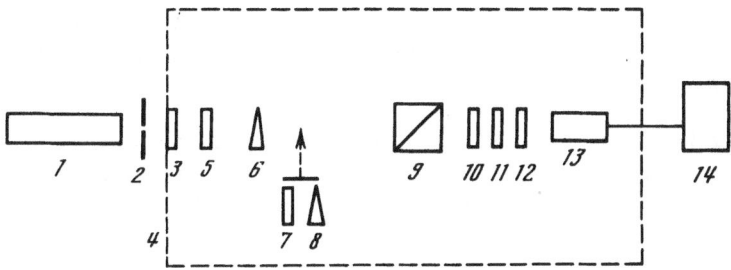

Fig. 7. Diagram of experimental setup for measuring the tensor components $\chi^{2\omega}$: 1) aluminum-yttrium garnet (containing neodymium) laser of type LT1PCh-7 ($f_{cycle} = 12.5$ Hz, $P_{peak} \sim 1$ MW $\cdot$ cm^{-2}; 2) diaphragm of diameter 2 mm; 3) KS-15 light filter; 4) opaque box; 5 and 12) neutral light filters; 6) wedgelike sample of crystal studied; 7) KS-15 light filter; 8) wedgelike sample of KDP; 9) polarizing Glan prism; 10) SZS-21 light filter; 11) interference light filter at $\lambda = 0.53$ μm; 13) photomultiplier of type FÉU-38; 14) oscilloscope of type S1-15.

TABLE 14[*]

ijk	$\left|\chi^{2\omega}_{ijk}\right|/\chi^{2\omega}_{zxy}$(KDP)		$\left|\Delta^{2\omega}_{ijk}\right|/\Delta^{2\omega}_{zxy}$(KDP)	
	m-Dinitro-benzene	m-Toluene-diamine	m-Dinitro-benzene	m-Toluene-diamine
xxz	2.5 ± 0.5 <2 [43]	1.0 ± 0.2	1.0 ± 0.2 <0.8 [43]	0.4 ± 0.1
yyz	4.8 ± 0.9 3.6 ± 1.8 [43]	2.3 ± 0.5	3.5 ± 0.6 2.6 ± 1.3 [43]	0.4 ± 0.1
zxx	3.0 ± 0.5 <2.5 [43]	0.9 ± 0.2	1.2 ± 0.2 <1.1 [43]	0.3 ± 0.1
zyy	6.2 ± 1.1 4.5 ± 2.2 [43]	2.6 ± 0.5	3.0 ± 0.5 2.1 ± 1 [43]	0.5 ± 0.1
zzz	1.7 ± 0.4 1.6 ± 0.8 [43]	1.8 ± 0.4	1.8 ± 0.4 1.8 ± 0.9 [43]	0.4 ± 0.1

[*] $\chi^{2\omega}_{zxy}$(KDP) $= +1.04\cdot10^{-9}$ cgse, $\Delta^{2\omega}_{zxy}$(KDP) $= +1.18\cdot10^{-9}$ cgse [46, 47].

$\chi^{2\omega}_{36} = 2\chi^{2\omega}_{312} \equiv 2\chi^{2\omega}_{zxy}$ [1]. It is seen from Table 14 that for both of the crystals we studied, the additional symmetry conditions of Kleinman [111] are satisfied (to without the accuracy of the measurements), as was to be expected.

§4. Second-Harmonic Generation in Crystals of m-Dinitrobenzene and m-Toluenediamine under Conditions of Vector Synchronism of Light Waves

As is well known [1], efficient doubling by nonlinear crystals of the frequency of optical radiation requires equality of the phases of the quadratic polarization wave and the second-harmonic light wave. This phase synchronism condition leads to the equation

$$N_1\mathbf{n}_1 + N_2\mathbf{n}_2 = 2N_3\mathbf{n}_3,$$ (1)

where the indices 1 and 2 label the wave normals $\mathbf{n}$ and corresponding indices of refraction N refer to the main radiation, while 3 refers to the second harmonic.

Because crystals of m-dinitrobenzene and m-toluenediamine are optically biaxial, like most crystals of organic compounds, it is of interest to consider phase synchronism in optically biaxial crystals.

In biaxial crystals, to each direction of the wave normal $\mathbf{n} = \{\sin\theta\cos\varphi,\ \sin\theta\sin\varphi,\ \cos\theta\}$ there correspond two mutually perpendicularly polarized planes of the wave, whose indices of refraction can be found from the Fresnel equation and have the form

$$N^{\pm} = \left[\frac{1}{2}\left(p \pm \sqrt{p^2 - 4q}\right)\right]^{-\frac{1}{2}}.$$ (2)

The quantities p and q are determined by the principal values of the indices of refraction N_x, N_y, N_z and the polar angles θ, φ:

$$p = (N_y^{-2} + N_z^{-2})\sin^2\theta\cos^2\varphi + (N_x^{-2} + N_z^{-2})\sin^2\theta\sin^2\varphi + (N_x^{-2} + N_y^{-2})\cos^2\theta,$$ (3)

$$q = N_y^{-2}N_z^{-2}\sin^2\theta\cos^2\varphi + N_x^{-2}N_z^{-2}\sin^2\theta\sin^2\varphi + N_x^{-2}N_y^{-2}\cos^2\theta.$$ (4)

In the general case, Eq. (1) expresses noncollinear, or vector, phase synchronism of the light waves. Such synchronism can be observed, e.g., if a collimated laser beam is partly scattered by inhomogeneities in a crystal* The nonlinear interaction of the oncoming radiation (n_1) with the scattered radiation (n_2) leads to most efficient generation of the second harmonic along the directions of synchronism (n_3) satisfying condition (1). Writing (1) in terms of the projections onto the principal dielectric axes X, Y, and Z (we ignore the dispersion of these axes), we obtain after some simple transformations the equation [113]

$$[N_{x1}^{-2}(A_y^2 + A_z^2) + N_{y1}^{-2}(A_x^2 + A_z^2) + N_{z1}^{-2}(A_x^2 + A_y^2)]$$
$$+ \xi^\pm \{[N_{x1}^{-2}(A_y^2 + A_z^2) + N_{y1}^{-2}(A_x^2 + A_z^2) + N_{z1}^{-2}(A_x^2 + A_y^2)]^2$$
$$- 4(A_x^2 + A_y^2 + A_z^2)(N_{y1}^{-2}N_{z1}^{-2}A_x^2 + N_{x1}^{-2}N_{z1}^{-2}A_y^2 + N_{x1}^{-2}N_{y1}^{-2}A_z^2)\}^{1/2} - 2 = 0 \tag{5}$$

with the parameters

$$A_x = 2N_3 \sin\theta_3 \cos\varphi_3 - N_1 \sin\theta_1 \cos\varphi_1,$$
$$A_y = 2N_3 \sin\theta_3 \sin\varphi_3 - N_1 \sin\theta_1 \sin\varphi_1,$$
$$A_z = 2N_3 \cos\theta_3 - N_1 \cos\theta_1,$$
$$\xi^\pm = \pm 1.$$

For given angles θ_1 and φ_1, Eq. (5) corresponds to the conical surfaces $\theta_3 = f(\varphi_3)$ formed by the set of directions of synchronous generation of the second harmonic. The sign over the quantity ξ in front of the curly brackets in (5) depends on the polarization of the scattered wave of the main radiation (which corresponds to the sign of N_2), and N_1 and N_3 are expressed by equations of type (2). Since $N^+ \leq N^-$, and $N_1 + N_2 > 2N_3$, for noncollinear phase synchronism, in the region of normal dispersion ($N_{1,2} < N_3$), finding $\theta_3 = f(\varphi_3)$ reduces to solving three equations of type (5) having the form

$$F[N_1^-(\theta_1, \varphi_1); \quad \xi^-; \quad N_3^+(\theta_3, \varphi_3); \theta_1; \varphi_1; \theta_3; \varphi_3] = 0, \tag{6}$$

$$F[N_1^-(\theta_1, \varphi_1); \quad \xi^+; \quad N_3^+(\theta_3, \varphi_3); \theta_1; \varphi_1; \theta_3; \varphi_3] = 0, \tag{7}$$

$$F[N_1^+(\theta_1, \varphi_1); \quad \xi^-; \quad N_3^+(\theta_3, \varphi_3); \theta_1; \varphi_1; \theta_3; \varphi_3] = 0. \tag{8}$$

Expression (6) determines the directions of synchronous generation of the second harmonic when waves of the main frequency with the same polarization interact (interaction of type I), while Eqs. (7) and (8) determine the directions of synchronism when the waves are orthogonally polarized (interaction of type II). For small birefringence $|N^+ - N^-|$, the conical surfaces (7) and (8) almost coincide. In the general case all three cones (6)-(8) of vector synchronism should be observed in crystals. The directions $n_2(\theta_2, \varphi_2)$ can be found from the equations

$$\tan^2\theta_2 = A_z^{-2}(A_x^2 + A_y^2); \quad \tan^2\varphi_2 = \frac{A_y^2}{A_x^2}.$$

For given angles θ_1, φ_1, and θ_2, φ_2, the direction of propagation of the second-harmonic wave, $n_3(\theta_3, \varphi_3)$, is determined by the equations

$$\tan^2\theta_3 = B_z^{-2}(B_x^2 + B_y^2),$$
$$\tan^2\varphi_3 = \frac{B_y^2}{B_x^2},$$

---

*Vector synchronism can also be observed during interaction in a crystal of the beams obtained by splitting (scattering) of the laser beam in front of the crystal.

where

$$B_x = N_1 \sin \theta_1 \cos \varphi_1 + N_2 \sin \theta_2 \cos \varphi_2,$$
$$B_y = N_1 \sin \theta_1 \sin \varphi_1 + N_2 \sin \theta_2 \sin \varphi_2,$$
$$B_z = N_1 \cos \theta_1 + N_2 \cos \theta_2.$$

In the case of a uniaxial ($N_x = N_y = N_o$, $N_z = N_e$) positive ($N_o < N_e$) (or negative, $N_e < N_o$) crystal, it follows from (2)-(4) that

$$N^+ = N_o, \quad N^- = N_e(\theta) \equiv [N_o^{-2} + (N_e^{-2} - N_o^{-2}) \sin^2 \theta]^{-\frac{1}{2}}$$
$$(N^+ = N_e(\theta), \ N^- = N_o) \tag{9}$$

and (5) becomes Eq. (17)* in [114]. In particular, for an interaction of the type (o,o) → e in an optically negative crystal (e.g., in a KDP crystal) the cones of emission of the second harmonic are given by

$$\sin^2 \varphi_3 = \frac{1}{2} N_{o1}^{-2} \sin^{-2} \theta_1 \sin^{-2} \theta_3 \{4N_{o1} [N_{o3}^{-2} + (N_{e3}^{-2} - N_{o3}^{-2}) \sin^2 \theta_3]^{-1/2}$$
$$\times \cos \theta_1 \cos \theta_3 - 2[N_{o3}^{-2} + (N_{e3}^{-2} - N_{o3}^{-2}) \sin^2 \theta_3]^{-1} - N_{o1}^2 (\cos 2\theta_1 + \cos 2\theta_3)\}.$$

If the wave normal lies in the plane (i, k) of the optical axes of the crystal, $\widehat{in} = \alpha$, $N_k < N_j < N_i$ and $V = \arcsin \left(\dfrac{N_k^{-2} - N_j^{-2}}{N_k^{-2} - N_i^{-2}} \right)^{1/2}$, then in the first quadrant

for $\alpha < V$

$$N^+ = [N_k^{-2} + (N_i^{-2} - N_k^{-2}) \sin^2 \alpha]^{-\frac{1}{2}}, \qquad N^- = N_j, \tag{10}$$

for $\alpha = V$

$$N^+ = N^- = N_j,$$

for $\alpha > V$

$$N^+ = N_j, \qquad N^- = [N_k^{-2} + (N_i^{-2} - N_k^{-2}) \sin^2 \alpha]^{-\frac{1}{2}}. \tag{11}$$

When a wave propagates in two different planes (kj) and (ji), N^+ and N^- are expressed by equations of the form (10) and (11), respectively.† Thus, the wave normals in the principal planes of a biaxial crystal correspond to indices of refraction which are determined in complete analogy with the indices of refraction of ordinary (o) and extraordinary (e) waves of a uniaxial crystal (9). Using this analogy, we will as in [115] denote a wave polarized perpendicular to the principal plane (ij) by the index o: $N_o \equiv N_k$, and a wave polarized parallel to the (ij) plane by e: $N_e(\alpha) \equiv [N_j^{-2} + (N_i^{-2} - N_j^{-2}) \sin^2 \alpha]^{-1/2}$, $\alpha = \widehat{in}$.

For the (ij) plane Eq. (15) splits into two equations,

$$(A_i^2 + A_j^2) N_{k1}^{-2} - 1 = 0 \quad \text{(scattered o-type wave)},$$
$$A_i^2 N_{j1}^{-2} + A_j^2 N_{i1}^{-2} - 1 = 0 \quad \text{(scattered e-type wave)},$$

---

*The factor r_1 in the denominator of the third fraction in square brackets is omitted in Eq. (17) of [114].

†When $N_i < N_j < N_k$ the expressions for N^+ and N^- are switched. The first index in the notation for the principal planes corresponds to the polar coordinate axis.

and the directions of propagation of the wave n_3 (α_3) of the second harmonic can be found from the equations

$$N_{e3}\,(\alpha_3) - N_{k1}\cos(\alpha_3 - \alpha_1) = 0 \tag{12}$$

for interaction of type (o,o) → e;

$$4N_3^2 - 4N_3 N_1 \cos(\alpha_3 - \alpha_1) + N_1^2 - N_{k1}^2 = 0 \tag{13}$$

for interactions of type (e, o) → o ($N_1 = N_{e1}\,(\alpha_1)$, $N_3 = N_{k3}$) or (e, o) → e ($N_1 = N_{e1}\,(\alpha_1)$, $N_3 = N_{e3}\,(\alpha_3)$);

$$(N_{i1}^{-2} - N_j^{-2})(2N_3 \sin\alpha_3 - N_1 \sin\alpha_1)^2 + N_{j1}^{-2}\,[4N_3^2 - 4N_3 N_1 \cos(\alpha_3 - \alpha_1) + N_1^2] - 1 = 0 \tag{14}$$

for interactions of type (o, e) → o ($N_1 = N_{k1}$, $N_3 = N_{k3}$), (o, e) → e ($N_1 = N_{k1}$, $N_3 = N_{e3}\,(\alpha_3)$) or (e, e) → o ($N_1 = N_{e1}\,(\alpha_1)$, $N_3 = N_{k3}$).

Phase synchronism is not possible in the region of normal dispersion for interactions of type (o, o) → o and (e, e) → e. The angles α_1 and α_3 are related to the angle

$$\alpha_2 = \arctan \frac{2N_3 \sin\alpha_3 - N_1 \sin\alpha_1}{2N_3 \cos\alpha_3 - N_1 \cos\alpha_1}.$$

If the angles α_1 and α_2 are known, then

$$\alpha_3 = \arctan \frac{N_1 \sin\alpha_1 + N_2 \sin\alpha_2}{N_1 \cos\alpha_1 + N_2 \cos\alpha_2}.$$

Equations (12)-(14) make it possible to find the apex angles of the cones of vector synchronism (5) in the principal planes of the crystal. Thus, for $\alpha_1 = 90°$ ($n_1 \parallel$ j) the apex angles of the cones of the directions $\Omega = \alpha_3^{(2)} - \alpha_3^{(1)}$ in the plane (ij) are equal to

$$2\arctan \left\{ \frac{[N_{j3}^{-4} + 4N_{k1}^{-2}(N_{i3}^{-2} - N_{j3}^{-2})]^{1/2} - N_{j3}^{-2}}{2\,(N_{i3}^{-2} - N_{j3}^{-2})} \right\}^{1/2} \tag{15}$$

for an interaction of type (o,o) → e;

$$2\arctan \left\{ \frac{4N_{k3}^2 + N_{i1}^2 - N_{k1}^2}{4N_{k3}N_{i1}} \right\} \tag{16}$$

for an interaction of type (e, o) → o;

$$2\arccos \{[N_{k1}N_{i1}^{-2} - N_{j1}^{-1}(4N_{j1}^{-2}N_{k3}^2 - 4N_{i1}^{-2}N_{k3}^2 + 2N_{j1}^{-2}N_{k1}^2 + N_{i1}^{-2}N_{j1}^2 - N_{i1}^{-2}N_{k1}^2 - 1)^{1/2}][2N_{k3}(N_{i1}^{-2} - N_{j1}^{-2})]^{-1}\} \tag{17}$$

for an interaction of type (o, e) → o.

The results obtained above make it possible to calculate all the directions of the wave normals of the main and transformed radiation at which noncollinear phase synchronism occurs in crystals of m-dinitrobenzene and m-toluenediamine (as well as other crystals).

Figure 8 shows the directions of propagation found by computer from (6)-(8) of the second harmonic due to synchronous nonlinear interaction in the crystal of m-dinitrobenzene of the oncoming and scattered laser radiation ($\lambda_1 = 1.064$ μm). It is seen that the vertex angles of the cones of the directions of propagation of the second-harmonic waves for a type I interaction are larger than for a type II interaction. Thus, when $n_1 \parallel$ x, $\Omega = 45°44'$ in the ZX plane for an

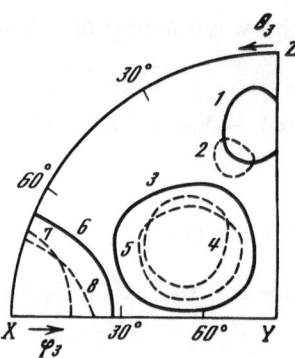

Fig. 8. Directions of wave propagation (θ_3, φ_3) of the second harmonic in the case of mixing of oncoming (θ_1, φ_1) and scattered laser radiation with $\lambda = 1.064\ \mu m$ under conditions of vector synchronism in crystals of m-dinitrobenzene (a quadrant of the stereographic projection is shown). 1) $\theta_1 = 15°$, $\varphi_1 = 60°$, interaction of type $(-, -) \to +$; 2) $\theta_1 = 30°$, $\varphi_1 = 60°$, interaction of type $(-, +) \to +$; 3) $\theta_1 = 60°$, $\varphi_1 = 50°$, interaction of type $(-, -) \to +$; 4) $\theta_1 = 60°$, $\varphi_1 = 50°$, interaction of type $(-, +) \to +$; 5) $\theta_1 = 60°$, $\varphi_1 = 50°$ interaction of type $(+, -) \to +$; 6) $\theta_1 = 90°$, $\varphi_1 = 0°$, interaction of type $(-, -) \to +$; 7) $\theta_1 = 90°$, $\varphi_1 = 0°$, interaction of type $(-, +) \to +$; 8) $\theta_1 = 90°$, $\varphi_1 = 0°$, interaction of type $(+, -) \to +$.

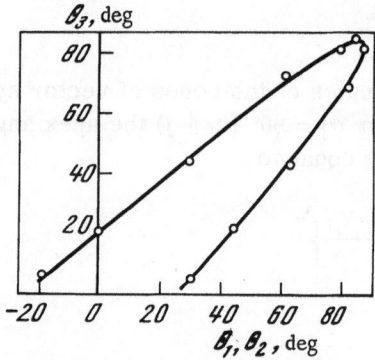

Fig. 9. Relation among the angles of vector synchronism for type $(o, o) \to e$ interaction of light waves in the ZX plane of crystals of m-toluenediamine for generation of the second harmonic with $\lambda = 0.53\ \mu m$. Points show experimental values of the angles.

interaction of type $(o, o) \to e$, while for an interaction of type $(e, o) \to e$, $\Omega = 32°16'$ in the same plane.

Relation (12) illustrated in Fig. 9 holds for crystals of m-toluenediamine for interaction of the type $(o, o) \to e$ in the ZX plane between the angles of vector synchronism, while for an interaction of type $(o, e) \to o$, in accordance with the solution of Eq. (14) there should be synchronous generation in the ZY plane of the second harmonic along the Y axis $(\theta_3 = 90°)$ at $\theta_1^{(1)} \approx 77°$, $\theta_2^{(1)} \approx 102°$, or at $\theta_1^{(2)} \approx 103°$, $\theta_2^{(2)} \approx 78°$.

In order to observe noncollinear interactions of light waves during generation of the second harmonic in crystals of m-dinitrobenzene and m-toluenediamine,* the setup shown in

*We first reported noncollinear synchronism in crystals of m-dinitrobenzene in [63]. In [50, 52, 55, 116] experimental data are cited for vector interaction in crystals of m-nitroaniline (without calculating the angles of synchronism).

Fig. 10. Diagram of experimental setup for
observing noncollinear interactions of light
waves in crystals: 1) neodymium-glass or
type LTIPCh-7 laser; 2) diaphragm;
3) KS-15 light filter; 4) mirror (50% transmission); 5) mirror (100% transmission);
6) crystal; 7) screen.

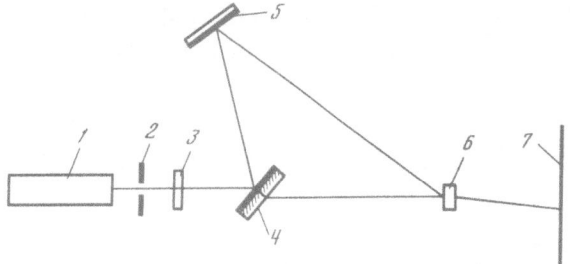

Fig. 10 was used. Light from a neodymium glass laser or type LTIPCh-7 laser 1 passed
through a diaphragm 2 and IKS-1 light filter 3, and was split into two parts by the 50% mirror
4. With the aid of the 100% mirror 5, both beams were brought together at a certain angle in
the crystal 6, which was placed on a rotating stand (the accuracy in determining the angles of
rotation was 1'. A screen 7 was placed at a distance 10-20 cm behind the crystal. Polarizing
Glan prisms (not shown in Fig. 10) could be placed between each mirror and the crystal as well
as between the crystal and the screen. Refraction of light rays by the surface of the crystal was
taken into account when measuring the angles θ_i and φ_i (i = 1, 2, 3).

The experimental data are in good agreement with the calculated values of the angles.
Thus, for nonlinear interactions of the type (o, o) → e in the ZX plane of crystals of m-dinitrobenzene, the vertex angles Ω of the second-harmonic cones increase from $\Omega \simeq 0$ to the maximal
values determined by equations of the form (15) when θ_1 changes from $\theta_1 \lesssim \theta_m$ to $\theta_1 = 90°$. The
intensity distribution of the second harmonic observed in the plane of the screen is shown in
Fig. 11 for certain cases of nonlinear synchronism in crystals of m-dinitrobenzene and m-toluenediamine. In the case of depolarized laser light, when various types of interaction of the light·
waves are occurring in the crystals, as well as when any one of a number of efficient scattering centers are present in the crystals (bubbles, cleavages, etc.), several second-harmonic
cones are registered. Nonuniformity of the intensity distribution of the second harmonic in the

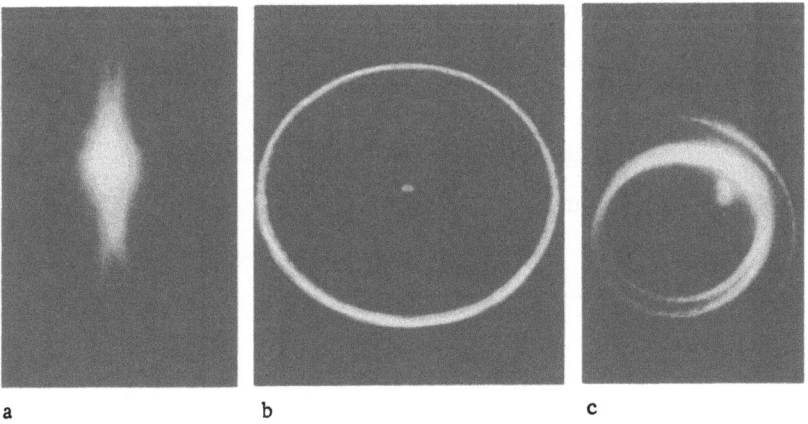

a   b   c

Fig. 11. Distribution of second-harmonic emission with $\lambda = 0.53$
μm in the near zone during noncollinear synchronism in crystals
of m-toluenediamine (a) and m-dinitrobenzene (b, c): a) (o, e) →
o interaction in the bc plane involving laser beams with $\widehat{n_1 b} = 13°$
and $\widehat{n_2 b} = -12°$; b, c) interaction of oncoming and scattered waves:
$n_1 \parallel a$, $e_1 \parallel b$; $\theta_1 = 60°$, $\varphi_1 = 30°$, for depolarized laser light.

plane of the screen is caused by the difference of the reflection coefficients from the output face of the crystal of second-harmonic rays incident at different angles on the output face of the crystal.

§5. Frequency Doubling of Laser Light in Crystals of m-Dinitrobenzene and m-Toluenediamine during Collinear Phase Synchronism

In a biaxial crystal $(n_1 \parallel n_2 \parallel n_3)$, Eq. (1) becomes the equality $N_1 + N_2 = 2N_3$ when there is collinear phase synchronism, and since $N^+ \leq N^-$, the directions in which the phases θ_m, φ_m of the collinearly propagating light waves add constructively in the region of normal dispersion $(N_3 > N_{1,2})$ are determined by the intersection (or tangency) of the inner nappe of the double surface $N_3^+(\theta, \varphi)$ of the indices of refraction at the frequency of the harmonic with the surfaces $N_1^-(\theta, \varphi)$ or $\frac{1}{2}[N_1^+(\theta, \varphi) + N_2^-(\theta, \varphi)]$:*

$$N_1^- = N_3^+,$$
$$N_1^+ + N_1^- = 2N_3^+,$$

or

$$p_2 + \sqrt{p_2^2 - 4q_2} + \sqrt{p_1^2 - 4q_1} - p_1 = 0 \tag{18}$$

for an interaction of type I,

$$2q_2^{-1}(p_2 - \sqrt{p_2^2 - 4q_2}) - q_1^{-1}(p_1 + 2\sqrt{q_1}) = 0 \tag{19}$$

for an interaction of type II, where p and q are expressed by Eqs. (3) and (4).† These equations describe the conic surfaces of the directions of synchronous generation of the second harmonic, $\theta_m = f(\varphi_m)$, when waves of the fundamental frequency interact which have the identical (18) or mutually orthogonal (19) polarizations. Thirteen different cases are possible for the directions of collinear synchronism in crystals [53].

If the wave normals of the interacting waves lie in the principal plane (i, j) of a biaxial crystal, then Eqs. (12)-(14) of the previous section imply the following formulas for the angles of synchronism:

$$\sin \alpha_m = \left(\frac{N_{k1}^{-2} - N_{j2}^{-2}}{N_{i2}^{-2} - N_{j2}^{-2}}\right)^{1/2} \tag{20}$$

for an interaction of type (o, o) → e;

$$\sin \alpha_m = \left(\frac{N_{k2}^{-2} - N_{j1}^{-2}}{N_{i1}^{-2} - N_{j1}^{-2}}\right)^{1/2} \tag{21}$$

---

*Interaction (tangency) of these surfaces corresponds to so-called critical (noncritical) phase synchronism.

†Here and below we will again use the notation of the preceding chapters, in which the index 1 refers to laser emission, and 2 to emission of the second harmonic.

for an interaction of type (e, e) → o;

$$\sin \alpha_m = \left[\frac{N_{j1}^{-2} - (2N_{k2} - N_{k1})^{-2}}{N_{j1}^{-2} - N_{i1}^{-2}} \right]^{1/2} \approx \left(\frac{2N_{k2} - N_{k1} - N_{j1}}{N_{i1} - N_{j1}} \right)^{1/2} \tag{22}$$

for an interaction of type (o, e) → o   or   (e, o) → o;

$$\sin \alpha_m \approx \left(\frac{2N_{j2} - N_{j1} - N_{k1}}{2N_{j2} - N_{j1} + N_{i1} - 2N_{i2}} \right)^{1/2} \tag{23}$$

for an interaction of type (o, e) → e or (e, o) → e.

The approximate equalities (22) and (23) are valid under the condition $|N_j - N_i| \ll N_i$, $N_{k2} - N_{k1} \ll N_{k1}$, which practically always holds. Equations analogous to (20)-(23) were first obtained in [54].

It should be noted that noncritical collinear phase synchronism along the principal axes of biaxial crystals (such synchronism is forbidden in the crystal class 222 by symmetry) is most favorable because the angular widths are larger and large effective interaction lengths are assured for the nonlinearly interacting light waves.

The coefficients $d^{2\omega}$ of nonlinear interaction have a rather complicated form in the case of collinear synchronism in biaxial crystals. For instance, for a type I interaction in crystals of symmetry mm2 with $\chi_{xxz}^{2\omega} \approx \chi_{zxx}^{2\omega}$ and $\chi_{yyz}^{2\omega} \approx \chi_{zyy}^{2\omega}$ [117], we have

$$d^{2\omega} \approx (\chi_{zyy}^{2\omega} - \chi_{zxx}^{2\omega})(3 \sin^2 \delta - 1) \sin \theta_m \sin 2\varphi_m \cos \theta_m \cos \delta - 3(\chi_{zxx}^{2\omega} \cos^2 \varphi_m + \chi_{zyy}^{2\omega} \sin^2 \varphi_m) \sin \theta_m \cos^2 \theta_m \sin \delta \cos^2 \delta$$

$$+ (\chi_{zxx}^{2\omega} \sin^2 \varphi_m + \chi_{zyy}^{2\omega} \cos^2 \varphi_m) \sin \theta_m \sin \delta (3 \sin^2 \delta - 2) + \chi_{zzz}^{2\omega} \sin^3 \theta_m \sin \delta \cos^2 \delta,$$

where the angle δ is related to the angles θ_m, φ_m of synchronism and the acute angle 2V between the optical axes by the relation

$$\cot 2\delta = (\cot^2 V \sin^2 \theta_m - \cos^2 \theta_m \cos^2 \varphi_m + \sin^2 \varphi_m)(\cos \theta_m - \sin 2\varphi_m)^{-1}.$$

If the nonlinear interactions occur in the principal planes of a biaxial crystal then the expressions for $d^{2\omega}$ simplify greatly [115].

Table 15 gives the results of calculating the angles of collinear synchronism in the principal planes of m-dinitrobenzene and m-toluenediamine crystals by Eqs. (20)-(23) for second-harmonic generation with $\lambda_1 = 1.064$ and 1.153 μm. In order to find the angles of phase synchronism θ_m outside the principal planes of the crystals it was necessary to solve the general equations (18) and (19). Figure 12 shows a quadrant of the stereographic projection of all the

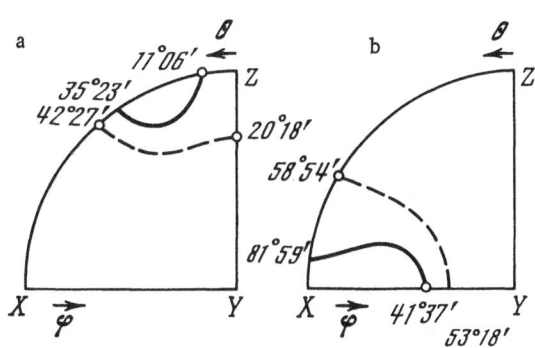

Fig. 12. Directions of collinear synchronism in crystals of m-dinitrobenzene (a) and m-toluenediamine (b) for second-harmonic generation with λ = 532 nm in which waves of the main beam with the same (solid line) and orthogonal (dashed line) polarizations interact.

TABLE 15*

Principal plane	Type of inter- action	Crystal			
		m-dinitrobenzene		m-toluenediamine	
		$\lambda_1{=}1.064\ \mu m$	$\lambda_1{=}1.153\ \mu m$	$\lambda_1{=}1.064\ \mu m$	$\lambda_1{=}1.153\ \mu m$
XY	$(o, o) \rightarrow e$ *	—	—	41°37′	39°33′
	$(e, e) \rightarrow o$	—	—	—	7°25′
	$(o, e) \rightarrow e$	—	—	53°18′	49°50′
ZX	$(o, o) \rightarrow e$	35°23′	34°04′ (35°46′)	81°59′	—
	$(e, e) \rightarrow o$ *	11°06′	14°30′ (15°50′)	—	—
	$(o, e) \rightarrow e$ *	42°27′	40°17′ (42°20′)	58°54′	64°18′
ZY	$(o, e) \rightarrow e$ *	20°18′	15°12′ (14°30′)	—	—

*The angles of synchronism for m-dinitrobenzene calculated in [43] from the indices of refraction are given in parentheses. In the XY plane the angle of synchronism is mea- sured from the X axis, while in the ZX and ZY planes it is measured from the Z axis. An asterisk denotes interactions which are forbidden by symmetry; they are only ob- served when the main beam is extracted from the principal planes.

possible directions found by solving these equations of collinear synchronism in crystals of m-dinitrobenzene and m-toluenediamine for second-harmonic generation with $\lambda_1 = 1.064\ \mu m$. A characteristic feature of m-dinitrobenezene and m-toluenediamine crystals is the fact that both cones of directions of collinear synchronism (corresponding to interactions of the light waves of types I and II) contain the optical axes of the crystals. In this case the cones of the directions n_3 corresponding to a given type of interaction when there is vector synchronism can only lie outside the cones of collinear synchronism corresponding to the same type of interaction of the light waves.

Table 16 gives the parameters characterizing collinear phase synchronism for interac- tion of light waves of the type $(o, o) \rightarrow e$ in the ZX plane during generation of the second harmonic of laser light with $\lambda_1 = 1.064\ \mu m$ by crystals of m-dinitrobenzene and m-toluene- diamine at room temperature. Here $\delta\theta_m$ and $\delta\lambda_1$ are the angular and spectral halfwidths of synchronism (in air); L is the thickness of the crystal in the direction of synchronism; L_a and L_q are the aperture and quasistatic lengths; D is the diameter of the laser beam; τ is the duration of the laser pulse.

Figure 13 shows the dispersion of the angles of phase synchronism θ_m calculated in terms of the $N_i(\lambda)$ which occurs during generation of the second harmonic of radiation of frequency $\lambda_1 = 0.87-1.15\ \mu m$ in the ZX plane [interaction of type $(o, o) \rightarrow e$] in crystals of m- dinitrobenzene and m-toluenediamine. For $\lambda_1 \approx 1.11\ \mu m$ at room temperature, noncritical phase synchronism should occur in a crystal of m-toluenediamine along the a axis with the main radiation polarized along the b axis and the second harmonic along the c axis.

Fig. 13. Dispersion of the angle of syn- chronism for generation of the second harmonic for interaction of type $(o, o) \rightarrow e$ in the (010) plane of crystals of m-dinitro- benzene (a) and m-toluenediamine (b).

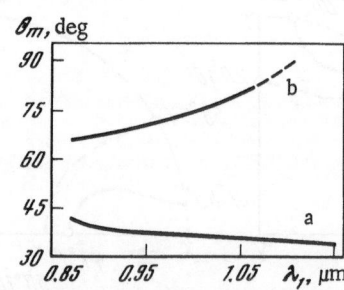

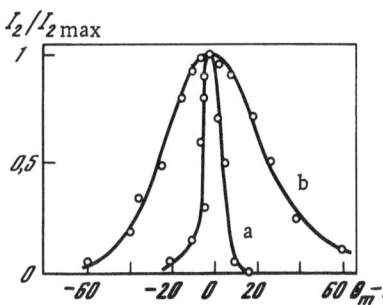

Fig. 14. Angular distribution (in air) of intensity of second-harmonic generation with λ = 532 nm for interaction of type (o, o) → e in the (010) plane of crystals of m-dinitrobenzene (a) and m-toluenediamine (b) of thickness 2 mm. The difference between θ and θ_m is given in angular minutes.

The parameters α_m and $\delta\alpha_m$ as well as the relative values of $d^{2\omega}$ and $\chi_{ijk}^{2\omega}$, which determine the synchronous generation of the second harmonic in the principal planes in crystals of m-dinitrobenzene and m-toluenediamine, were measured for radiation with $\lambda_1 = 1.064\ \mu$m on the setup described in Sec. 3 (cf. Fig. 7). The plane-parallel waferlike samples of thickness $L \simeq 2$ mm were attached to a rotating stand rotatable to within 1', and a diaphragm 2 was placed in the resonator to decrease the divergence of the laser beam. The error in measuring the angles of synchronism was mainly caused by inaccuracy in cutting the samples and did not exceed $\pm 1°$. Within the limits of this error, the results of the measurements are in good agreement with the calculated values of α_m (cf. Table 15).* It should be noted that the accuracy in measuring the angle of synchronism in the ZX plane for the crystal of m-toluenediamine was not worse than 5' due to the possibility of recording the second harmonic in the two directions θ_m and $\pi - \theta_m$† which are nearly the same. The observed dependence $I_2\ (\theta_m - \theta)$ for an (o, o) → e interaction in the ZX plane for crystals of m-dinitrobenzene and m-toluenediamine is shown in Fig. 14. Since the divergence of the laser beam does not exceed the $\delta\theta_m$ measured and the width of the laser emission spectrum is ~0.5 Å ≪ $\delta\lambda_1$, the increase of $\delta\theta_m^{\exp}$ by a factor of 1.5 to 2 as compared with $\delta\theta_m$ (cf. Fig. 14 and Table 16) can apparently be attributed to the mosaic (block) structure characteristic of molecular crystals.

The intensity I_2 of the second harmonic was recorded in the directions of synchronism, which made it possible (ignoring absorption and aperture effects under our experimental conditions) to find the relative values of $d^{2\omega}$ of the crystals studied,

$$\frac{|d^{2\omega}|}{d_{\text{KDP}}^{2\omega}} \simeq \left(\frac{I_2}{I_2^{\text{KDP}}}\right)^{1/2} \left(\frac{N_1 + 1}{N_1^{\text{KDP}} + 1}\right)^2 \frac{N_2 + 1}{N_2^{\text{KDP}} + 1} \frac{L^{\text{KDP}}}{L}.$$

The laser beam propagated in the direction $\theta_m' = 43°$, $\varphi_m' = 45°$ in the standard KDP crystal sample and interaction of light waves of the type (o, o) → e, occurred with coefficient

$$d_{\text{KDP}}^{2\omega} = \chi_{zxy}^{2\omega}(\text{KDP}) \sin\theta_m' \sin 2\varphi_m' \simeq 0.68\chi_{zxy}^{2\omega}(\text{KDP}).$$

The values of the coefficients of nonlinear interaction in crystals of m-dinitrobenzene (MDB) and m-toluenediamine (MTD)

$$|d_{\text{I}}^{2\omega}(\text{MDB})| = 4.2\chi_{zxy}^{2\omega}(\text{KDP}), \qquad |d_{\text{I}}^{2\omega}(\text{MTD})| = 2.5\chi_{zxy}^{2\omega}(\text{KDP}),$$
$$|d_{\text{II}}^{2\omega}(\text{MTD})| = 1.3\chi_{zxy}^{2\omega}(\text{KDP}),$$

---

*We did not measure α_m for $\lambda_1 = 1.153\ \mu$m. However, using the data of [41], $\theta_m^{\exp} = 34°45' \pm 30'$ at this wavelength for m-dinitrobenzene, which coincides with our value $\theta_m = 34°04' \pm 20'$.

†The experimental value for this angle of synchronism is $81°40' \pm 5'$.

TABLE 16

Crystal	θ_m	$\dfrac{d\theta_m}{d\lambda_1}\Big|_{\lambda_1=1.064\ \mu m}$, min/Å	$\delta\theta_m \cdot L$, min · cm	$\delta\lambda_1 L$, Å · cm	L_a/D	L_q/τ, cm/psec
m-Dinitrobenzene	35°23′	−0.1	1	10	6.2	0.45
m-Toluenediamine	81°59′	0.7	7	10	45	30

measured with accuracy 15-20% and the data of the table in [115] imply, assuming the additional Kleinman symmetry conditions hold, that

$$|\chi^{2\omega}_{yyz}\ (\text{MDB})\,| \simeq |\chi^{2\omega}_{zyy}\ (\text{MDB})\,| = (6.0 \pm 1.0)\,\chi^{2\omega}_{zxy}\ \text{KDP}$$
$$|\chi^{2\omega}_{yyz}\ (\text{MTD})\,| \simeq |\chi^{2\omega}_{zyy}\ (\text{MTD})\,| = (2.5 \pm 0.4)\,\chi^{2\omega}_{zxy}\ \text{KDP}$$
$$|\chi^{2\omega}_{zxx}\ (\text{MTD})\,| \simeq |\chi^{2\omega}_{xxz}\ (\text{MTD})\,| = (1.0 \pm 0.2)\,\chi^{2\omega}_{zxy}\ \text{KDP}$$

These results are in good agreement with the values of $|\chi^{2\omega}_{ijk}|$ found previously using the wedge method (cf. Table 14). The values of $|\chi^{2\omega}_{ijk}|$ as well as α_m and $\delta\alpha_m$ obtained for different substances coincide within the errors of the measurements.

The setup shown in Fig. 15 was used to find the (energy) conversion factors for laser light with $\lambda_1 = 1.06\ \mu m$ into the second harmonic for an interaction of type (o, o) → e in crystals of m-dinitrobenzene and m-toluenediamine. A neodymium-glass laser with energy $E_1 \simeq 1$ J was used with a pulse of duration 40 nsec. Q-switching of the laser resonator by a rotating prism with total internal reflection caused the wave vector n_1 of the laser light to vary during the lasing process in the plane of rotation of the prism. Therefore, in order to eliminate the effect of fluctuations of n_1 on $\eta = E_2/E_1$, the principal plane (010) of the nonlinear crystals, in which interaction of the type (o, o) → e of light waves was observed, was positioned perpendicular to the plane of rotation of the prism. The condition for optimal focusing [118, 119] was satisfied inside the crystals. In order to decrease losses due to reflection, the prism 10 was located at the position of minimal deflection.

As was shown by the experiments, volume destruction of the crystals studied starts to manifest itself at power densities of laser emission $P_1 \sim 300$ MW · cm^{-2}, and surface destruction appears at $P_1 \sim 200\text{-}250$ MW · cm^{-2} and depends strongly on the quality of treatment of the

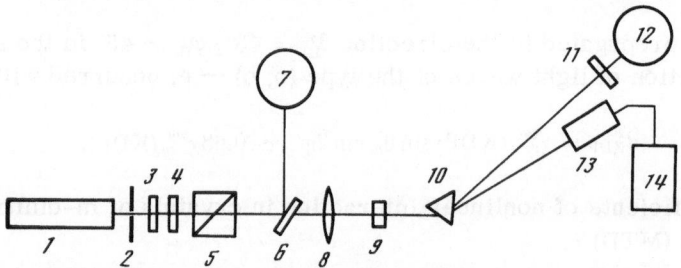

Fig. 15. Diagram of experimental setup for determining the conversion factor of laser light with $\lambda = 1.06\ \mu m$ into the second harmonic. 1) Neodymium laser; 2) diaphragm, diameter 5 mm; 3) KS-15 light filter; 4) neutral light filters; 5) polarizing Glan prism; 6) glass slab; 7 and 12) calorimeters; 8) lens; 9) nonlinear crystal; 10) glass prism; 11) SZS-21 light filter; 13) FÉK-15; 14) type 12-7 oscilloscope.

Fig. 16. Experimental dependence of the conversion factor
(with respect to energy) of
laser light into the second-harmonic on the power density of
the neodymium laser for crystals of m-dinitrobenzene (DNB)
with L ≃ 3 mm (a) and m-toluenediamine (MTD) with L ≃ 10
mm (b).

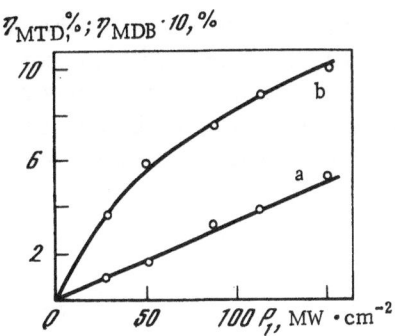

sample surfaces. The dependence $\eta(P_1)$ for both crystals obtained at $P_1 \lesssim 150$ MW·cm^{-2} is shown in Fig. 16. Pronounced aperture effects and the moderate values of the emission power density causing surface destruction of crystals of m-dinitrobenzene did not permit us to attain $\eta_{MDB} > 0.5\%$ under our experimental conditions, in spite of the large nonlinear interaction co-efficient ($d_{(MDB)}^{eoo} \simeq 6.2 d_{(KDP)}^{eoo}$) for these crystals. At the same time, for crystals of m-toluene-diamine with $d_{(MTD)}^{eoo} \simeq 3.7 d_{(KDP)}^{eoo}$ and L ≃ 10 mm a conversion coefficient of 10% was obtained for the second harmonic. When longer crystals are used and the width of the lasing spectrum is decreased, η_{MTD} can be substantially increased.

The critical sensitivity to angular detuning, the small aperture lengths, and the softness of crystals of m-dinitrobenzene which prevent qualitative treatment make it difficult to use these crystals in second-harmonic generators. However, the use of crystals on m-dinitro-benzene in studying scattering of light by polaritons [120] and dispersion of quadratic suscep-tibility [121] show that they can be of interest in investigations of the most diverse optical phenomena. On the other hand, the large optical nonlinearities, small aperture effects, and stability to destruction by high-power laser emission make crystals of m-toluenediamine promising as frequency doublers of pulsed laser light with λ = 1.06 μm.

CHAPTER 4

GENERATION OF THE SECOND OPTICAL HARMONIC
AND MOLECULAR AND CRYSTAL STRUCTURE
OF ORGANIC COMPOUNDS

§1. Nature of the Optical Nonlinearities of Organic Molecules

Viewing crystals of organic compounds as a collection of relatively weakly interacting molecules (cf. Chap. 1), one might expect that the linear and nonlinear optical effects of these crystals are mainly determined by two factors — the electronic structure of the individual molecules, and position in the unit cell.

We first discuss the effect of the structure of the constituent molecules of organic crys-tals characterized by different spatial orientation and symmetry of the electron orbitals (of σ-, π-, l-, or n-type) on the quadratic susceptibility of these crystals.

For the simplest bonding molecular orbital consisting of two weakly overlapping atomic orbitals $|A\rangle$ and $|B\rangle$ of the form $|\psi\rangle = (1 + \lambda^2)^{-1/2}(|A\rangle + \lambda|B\rangle)$, where λ is a hybridization parameter, the linear polarizability is $\alpha \sim (1 - \varepsilon^2)^2 d^4$, and the quadratic susceptibility is $\beta^{2\omega} \sim \varepsilon(1 - \varepsilon^2)^3 d^7 \sim \varepsilon d\alpha^{3/2}$ [122]. Here $\varepsilon = (1 - \lambda^2)/(1 + \lambda^2)$ is the extent to which the bond is ionic, d is the bond length. Since $e\varepsilon d$ is the constant dipole moment of the bond μ, the direction of which depends on λ, the sign of $\beta^{2\omega}$ is determined by the polarizability of the bond. For a purely ionic bond ($\lambda = 0$), when all the electrons are rigidly held by the anion and there is no redistribution of electron density caused by the external field, α and $\beta^{2\omega} \sim 0$. As the bond becomes less ionic, α and $\beta^{2\omega}$ increases, but for a purely covalent bond ($\lambda = 1$), when the valence electrons are in a field with an inversion center, $\beta^{2\omega}$ again vanishes by virtue of symmetry.

Thus, the quadratic polarizability is determined by the extendedness and asymmetry of the elctron orbitals [123]. This asymmetry can be due to a difference in the elctronegativities of the atoms or the sizes of the atomic orbitals, and also to hybridization of these orbitals or of the orbitals of unshared electron pairs, etc. A hybrid, or mixed, atomic orbital cannot have a center of inversion, and as a result of hybridization an unshared pair of electrons can contribute greatly to the polarizability as, e.g., in the case of hexamethylenetetramine [57, 58].

The combination of high electron mobilities (linear polarizabilities) with a highly asymmetric electron distribution, which is required for obtaining large optical nonlinearities, correlated well with Miller's phenomenological rule [112] $\chi^{2\omega}_{ijk} = \chi^{2\omega}_{ii}\,\chi^{\omega}_{jj}\,\chi^{\omega}_{kk}\Delta^{2\omega}_{ijk}$, where $\chi_{\xi\xi} = (4\pi)^{-1} \times (N^2_\xi - 1)$ are the linear susceptibilities, and $\Delta^{2\omega}_{ijk}$ is a coefficient characterizing the departure of the electron density distribution from a centrosymmetric one.* It also correlates with the anharmonic oscillator model [124]. This model takes into account the motion of an optical electron oscillator in a potential well defined by the interaction of the electron with the surrounding particles. For frequency-doubling, it is the deviation of this potential from a centrosymmetric potential, expressed in the general case by the 2^{2k+1}-pole moment of the charge distribution, that is important. In weak light fields, when the displacements of the electrons are small and the electron is located at the "bottom" of the well, the asymmetry of the potential has no effect on the motion of the electron. The electron is like a harmonic oscillator. In an intense laser field, the asymmetry of the potential starts to manifest itself, and the effect becomes stronger the larger the displacement of the electron and the polarizability of the oscillator. The quasi-elastic forces which arise here depend nonlinearly on the magnitude of the displacement. The response of the electron to the external light field becomes distorted, i.e., the optical electron behaves like an anharmonic oscillator. From a quantum mechanical point of view, when the self-consistent field in which the electron is present lacks central symmetry, the electron states lose their well-defined parity and the probability of transitions contributing to the nonlinear polarizability is nonzero. For an anharmonic oscillator, transitions with $|\Delta n| > 1$ become allowed.

In accordance with the above arguments, one should expect in the case of organic compounds large nonlinear polarizabilities $\beta^{2\omega}$ for molecules which have a markedly asymmetric cloud of (delocalized) π-electrons (which are the most "mobile"). Such a deformation of a π-cloud in a molecule occurs, e.g., when the π-orbitals of conjugated bonds interact with non-bonding l- and n-orbitals of heteroatoms. This interaction leads to a redistribution of the density of the π-electrons over a common molecular orbital among various groups of atoms, of which some play the role of electron donors and others the role of electron acceptors. Thus, with respect to aromatic hydrocarbons with π- (and σ-) orbitals and a uniform distribution of π-charge, the following groups of atoms with l-electrons are electron donors: $\ddot{\text{N}}(\text{CH}_3)_2$, $\ddot{\text{N}}\text{H}_2$, $\ddot{\text{O}}\text{H}$, etc., while the groups of atoms NO_2, COH, $\text{C}\equiv\text{N}$, etc.† with n-electrons are acceptors.

_____

*That the $\Delta^{2\omega}_{ijk}$ (in the long-wave approximation) are proportional to the corresponding components of the octupole moment of the electron density distribution in the ground state is shown in [57].

†Whether groups of atoms are donors or acceptors relative to one another is determined by the relative position of their occupied or unoccupied orbitals.

The effect on the quadratic polarizability of aromatic molecules due to substituents with heteroatoms can be used to explain the increase observed by us (Chap. 2) of the second-harmonic signal in going from powders of phenanthrene, 1,2-benzanthracene, triphenylmethane, and 3,4-benzpyrene, which have an almost homogeneous π-charge distribution, to the powders of resorcin, m-toluenediamine, m-dinitrobenzene, and m-nitroaniline, which have increasing anisotropy of the π-electron density in a molecule.

One of the most accessible parameters characterizing the departure from centrosymmetry of the electron density in a molecule is the molecular dipole moment μ in the ground state. We pointed out for the first time in [63] the correlation of μ with optical nonlinearities in an analysis of the efficiency of second-harmonic generation in crystals of a series of disubstituted benzenes with donor and/or acceptor groups. Indeed, in this case, $\beta^{2\omega} \sim \varepsilon d \sim \mu$, in accordance with what was discussed above. Moreover, $\beta^{2\omega}$ increases when the donor-acceptor properties of the substituents increase, and with increasing distance d between them due to chain conjugation of the common π-orbitals. This last conjugation also leads to an increase in α. However, the experimentally measurable molecular dipole moment does not permit an estimation of the value of $\beta^{2\omega}$ for the system studied, since it contains not only a π- but also a σ-component, which introduces an uncertainty into the results. More accurate conclusions concerning the relation between $\beta^{2\omega}$ and μ can clearly be made in the case of saturated compounds with localized n- and σ-orbitals.

It should be noted that the vanishing of μ does not mean that the distribution of donor or acceptor substituents perturbing the electron density distribution in the hydrocarbon is centrosymmetric. Indeed, it is possible for the vector sum of the dipole moments associated with such substituents to be zero without the molecule having a center of inversion and without all the components of the octupole moment being equal to zero. Similarly, in many crystals consisting of polar molecules but not possessing spontaneous polarization, there is no center of symmetry, and generation of the second harmonic is observed.

§2. Role of Configurations with Charge Transfer

in Generation of the Second Optical Harmonic

As is well known, the polarizability of a quantum mechanical system is determined by the set of its energy levels and the transitions between them. It follows from the general expression* for the quadratic polarizability $\beta^{2\omega}$ that in the absence of resonance between the eigenfrequencies of the system ω_1, ω_2, ... and the frequencies ω and 2ω of the main radiation and its second harmonic, we have e.g.,

$$\beta_{xxx}^{2\omega} = 3\hbar^{-2}e^3 \sum_{i=1}^{\infty} \sum_{k=1}^{\infty} \{A_i x_{0i}^2 (x_{ii} - x_{00}) + A_{ik} x_{0i} x_{ik} x_{k0}\}. \tag{24}$$

Here the wave functions are assumed real and $x_{ik} = x_{ki}$,

$$A_i = \omega_i^2 [(\omega_i^2 - \omega^2)(\omega_i^2 - 4\omega^2)]^{-1},$$

$$A_{ik} = 2 \{\omega_i \omega_k [\omega^2 (\omega_i \omega_k + 8\omega^2) + \omega_i^2 \omega_k^2] - \omega^2 (\omega_i^2 + \omega_k^2)(2\omega^2 + 3\omega_i \omega_k)\}$$
$$\times [(\omega_i^2 - \omega^2)(\omega_i^2 - 4\omega^2)(\omega_k^2 - \omega^2)(\omega_k^2 - 4\omega^2)]^{-1}. \tag{25}$$

It is seen from Eqs. (24) and (25) that the main contribution to $\beta^{2\omega}$ of the molecule comes from the intense electron transitions with frequencies close to 2ω and ω. Taking into account

---

*See, for example, [57, 125].

†When the levels 0, m, and n are equidistant and $x_{0n} \ll x_{0m}$, this equation implies the expression obtained in [126] for $\beta^{2\omega}$ for a one-dimensional quantum oscillator with weak anharmonicity.

only intermediate levels m and n associated with such transitions, we have

$$\beta_{xxx}^{2\omega} = 3\hbar^{-2}e^3 \{A_m x_{0m}^2 (x_{mm} - x_{00}) + A_{mn} x_{0m} x_{mn} x_{n0} + A_n x_{0n}^2 (x_{nn} - x_{00})\}^{48}, \qquad (26)$$

where A_m, A_{mn}, and A_n are calculated from equations of the form (25).

In molecular systems with a center of inversion not possessing accidental degeneracy, the electron wave functions $|0\rangle$, $|m\rangle$ and $|n\rangle$ have a definite parity. Hence in such systems,

$$x_{jj} = 0, \quad x_{0m} x_{mn} x_{n0} = 0 \quad \text{and} \quad \beta^{2\omega} = 0. \qquad (27)$$

However, values of $\beta^{2\omega} \simeq 0$ can be observed even in acentric molecules. Such a situation occurs, e.g., in noncentrosymmetric conjugated hydrocarbons in which the distribution of the π-electrons is uniform over all the atoms of the hydrocarbon [127]. In these systems, symmetrically located ("paired") energy levels of bonding and antibonding molecular π-orbitals are associated respectively [128] with states of electrons and "holes" having opposite parities, which leads to equalities of the type (27). The quadratic polarizability of substituted hydrocarbons can be explained [129] by the contribution to their electron states of configurations with charge transfer.

The concept of configurations with charge transfer is associated with the interpretation of real molecular systems as being composed of fragments. In this interpretation, the states of the system are mixed and are determined by the contributions of configurations responsible for the electron density distribution both in the individual fragments (atoms or groups of atoms) and among them. The electron density distribution among the parts of the system which appear as electron donors and acceptors is often called the "charge transfer," both in the ground and in excited states.

Configurations with charge transfer correspond to ionic structures in the well-known valence bond approach. Charge transfer is involved, e.g., in the transition of an H_2 molecule into the $^1\Sigma_u^+$ state [130], and also in the formation of hydrogen bonds [131]. The contribution of configurations with charge transfer to lower states of the system* which are the most important for second-harmonic generation is responsible for the quadratic polarizability of organic molecules.

Indeed, for the nitrobenzene molecule, for example, the states ψ_0, ψ_2, and ψ_5 are mixed [132], so that electron transitions among them are allowed, in contrast to the case for benzene. Moreover, the most intense transitions, and the ones closest in wavelength to the wavelength λ of the second harmonic of neodymium laser light, is the transition to the intermediate state 2 ($f_{02} = 0.48$; $\lambda_{02} \simeq 250$ nm), to which the main contribution is from a configuration with charge transfer from the ring to the nitro group, and also the transition to the intermediate state 5 ($f_{05} = 0.30$; $\lambda_{05} \simeq 180$ nm), to which the main contribution is from the electron configuration $A_{1g} \rightarrow B_{1u}$ of benzene [132-134]. Both transitions are polarized along a second-order axis of symmetry of the molecule. Since in nitrobenzene with m = 2, n = 5, j = 2.5$x_{mn} \simeq x_{0m}$ and $x_{jj} - x_{00} \simeq x_{0j}$ [134, 135], we have [since, from (25), $A_m : A_{mn} : A_n \simeq 0.7 : 1.0 : 0.4$] that the ratio of the terms in the curly brackets in (26) is approximately 1 : 1 : 0.2 for nitrobenzene at the wavelength 530 nm of the second harmonic.

Thus in the typical case considered, the first term in (26) does not dominate and the use [37] of a two-level scheme to describe the generation of the second harmonic in which only one

*In quantum mechanical calculations the wave functions of these states can be represented as a linear combination of singly excited configurations of a definite symmetry.

transition with charge transfer is chosen* is in general unjustified. However, the observation of such a transition in the absorption or emission spectra is a spectroscopic criterion for acceptor−donor interaction of the fragments of a molecule requiring allowance for the contributions of configurations with charge transfer.

Quantum mechanical calculations (see, e.g., [136]) show that when the ionization of potential of the donor fragment decreases and the electron affinity of the acceptor fragment increases, as well as when the chain of conjugation becomes longer, the contribution of configurations with charge transfer in the ground and excited electron states increases.† The role of substituents of various nature and of the chain length of conjugation in the optical properties of molecules was noted by us in [62].

With increasing contribution of configurations with charge transfer to the molecular electron states, the molecule becomes more "ionic" and the dipole moment increases, while the transition (band) with charge transfer is displaced toward long wavelengths. All this increases the asymmetry of the electron distribution in the molecule, its polarizability, and hence also (cf. Sec. 1 of this chapter) the quantity $\beta^{2\omega}$.

In molecules with σ-bonds only, contributions to the electron states of configurations with charge transfer

$$n \to \sigma^*$$

are important for the quadratic polarizability (this holds, e.g., for hexamethylenetetramine, formaldehyde, and similar compounds with heteroatoms), whereas in molecules with conjugated bonds, it is the configurations

$$(\pi l) \to \pi^* \to \pi^* (n) \tag{28}$$

(e.g., m-nitrophenol, m-nitroaniline, and other compounds containing the groups $N(CH_3)_2$, NH_2, OH, OCH_3, etc., which contain l-electrons, and the groups NO_2, COH, $COCH_3$, CN, etc., containing π- and n-electrons),

$$\pi \to \pi^* (n) \tag{29}$$

(e.g., m-dinitrobenzene, benzil, and other compounds with groups containing only σ-, π-, and n-electrons), and

$$(\pi l) \to \pi^* \tag{30}$$

(e.g., m-dihydroxybenzene, m-toluenediamine, and other compounds with groups containing only σ- and l-electrons) that are important.

For systems with conjugated bonds, we consider the contribution of the above configurations to be much greater than from the configurations $n \to \sigma^*$ (from energy considerations) and $n \to \pi^*$ (from symmetry considerations).

If the $\pi(p)$-orbitals of the fragments are not coplanar, the contribution to the electron states of configurations with charge transfer (28)-(30) drops [134]. This is in agreement with the small effect of frequency doubling of neodymium laser light in noncentroasymmetric crystals of 1,8-dinitronaphthalene, where steric hindrance leads to considerable rotation of the

---

*A transition with charge transfer is the usual name for a transition for which the contribution of configurations with charge transfer is predominant.

†The contribution of configurations with charge transfer in the ground state for molecules of nitrobenzene reaches 6.6% [132], and 40% for complexes of amines with iodine [137].

V. D. SHIGORIN

TABLE 17

R	$\Delta\lambda$, nm	$\left\|\beta_{333}^{2\omega}\right\| \cdot 10^{31}$, cgse $\cdot$ cm^3	R	$\Delta\lambda$, nm	$\left\|\beta_{333}^{2\omega}\right\| \cdot 10^{31}$, cgse $\cdot$ cm^3
CH$_3$	3	0.9	CN	20.5	6.3
Cl	6	1.8	NH$_2$	26.5	8.1
Br	6.5	2	COCH$_3$	42	13
OH	7.0	2.2	COH	46	14
OCH$_3$	13.5	4	NO$_2$	65	20

planes of the nitro groups relative to the plane of the benzene rings and to a decrease of the contributions to the electron states of the configurations

$$\pi \to \pi^* \, (n).$$

Let us consider the simplest and best-studied aromatic compounds, the monosubstituted benzenes with the general formula $\phi - R$.

The symmetry of the π-electron cloud of the benzene ring and the axial symmetry of excitation along the direction of the nearest atom in the R radical lead to the symmetry mm2 for the system of π-electrons of the molecule (to a rough approximation) and $\beta_{311}^{2\omega} \simeq \beta_{113}^{2\omega} = 0$, $\beta_{322}^{2\omega} \simeq \beta_{223}^{2\omega} = \beta_{333}^{2\omega}/3$ (the 1-axis is perpendicular to the plane of the phenyl ring, the 3-axis is parallel to the C−R bond) [138]. Because of this, the vector part of the quadratic molecular polarizability

$$\beta_v^{2\omega} = \sqrt{\frac{3}{5}}(\beta_{311}^{2\omega} + \beta_{322}^{2\omega} + \beta_{333}^{2\omega}),$$

proportional to the constant dipole moment μ_3 of the system [139], is approximately equal to to $\beta_{333}^{2\omega}$. This result agrees with the conclusions found in the previous section concerning the relation between $\beta^{2\omega}$ and the dipole moments.

Although reliable data on the π-dipole moments of $\phi - R$ compounds are not available, the associated shift to longer wavelengths of the 200 nm-band of benzene [140] can serve as a measure of the redistribution of the π-electron density along the 3-axis of the molecule when substituents of different types are introduced into the ring. This "shift" $\Delta\lambda$ in fact determines the position of the charge transfer band in $\phi - R$ (transition between electron orbitals symmetric under a 180° rotation about the second-order 3-axis) relative to the band $A_{1g} \to B_{1u}$ of benzene and lets us put $\left|\beta_{333}^{2\omega}\right| \sim \Delta\lambda$ * when estimating $\beta^{2\omega}$ in many monosubstituted benzenes.

Table 17 lists some benzene substituents, the changes $\Delta\lambda$ caused by them in [142], and the calculated values of $\left|\beta_{333}^{2\omega}\right|$. In our estimates we used the value $\left|\beta_v^{2\omega}\right| = 20 \cdot 10^{-31}$ cm^3 (cgse) taken from [143] for the nitrobenzene molecule.

Taking into account the electron donor and electron acceptor properties of the substituents and the directions of the displacements of the electron density caused by them in the ring,

---

*In our treatment we have refrained from using the so-called mesomeric moments, since first of all they do not always correspond to the true π-dipole moments [141] and second, the results from a determination of the mesomeric moments of different workers often differ greatly. Hence the recent calculation in [138] in terms of mesomeric moments of $\chi_v^{2\omega}$ for a whole series of organic crystals seems to us to be incorrect.

the values of β_{333}^2 are either positive or negative. Bearing in mind the effect of lengthening of the conjugated chain and the contributions of the electron configurations with charge transfer between the two radicals R_1 and R_2, still larger values of $\beta_{333}^{2\omega}$ should be expected for para-disubstituted benzenes p-R_1-ϕ-R_2 in which R_1 and R_2 differ chemically. Such compounds include p-nitrophenol (R_1 = OH, R_2 = NO_2), p-nitroaniline (R_1 = NH_2, R_2 = NO_2), etc.

The results obtained are in accordance with the calculations in [132, 134] of the contributions of configurations with charge transfer to the lower electron states of molecules of aniline, nitrobenzene, and p-nitroaniline, and indicate that the quadratic optical nonlinearities increase in the series of compounds with (σ, π), (σ, π, l), (σ, π, n), and (σ, π, l, n)-orbitals.

§3. Oscillator Model of a Molecular Crystal.

Crystal Structure and Quadratic Susceptibility

The theoretical determination of the nonlinear optical susceptibility $\chi^{2\omega}$ for crystals is a complicated problem, and various models have been applied in its solution. One of the methods of solution involves the use of a microscopic model of nonlinear polarization, which enables one to relate the components of $\chi^{2\omega}$ to the microstructure of the crystal. This approach is used, e.g., in the case of cubic molecular crystals of hexamethylenetetramine [58] and in semiconductors of type $A^{II} B^{VI}$ [144] in which the nonlinear polarizabilities of the σ-bonds are additive as tensors, as well as for ferroelectric crystals with bonds of ionic character [145]. When such calculations are carried out, the structures of the polarizable elements are as simple as possible — that is, they are an ion or a σ-bond, at which the optical electrons are localized.

For molecular crystals of organic compounds, the majority of which have branched conjugated bonds, delocalization and sharing of electrons between the bonds (by groups of atoms) greatly complicate the problem of calculating the nonlinear polarizabilities of the structural elements and allowing for the local (effective) field inside the crystal. However, it is sometimes possible just from the known crystal structure of a substance to obtain approximate relations between the $\chi_{ijk}^{2\omega}$ and even estimate $\beta^{2\omega}$, if certain pairs of molecular fragments R_q and R_t between which charge transfer takes place during the transitions responsible for second-harmonic generation are modeled by a one-dimensional anharmonic oscillator R_q-R_t [78].

Let a unit cell of volume V_0 of the crystal contain $\varkappa$ oscillators of the above type with macroscopic values of the quadratic polarizability given by

$$\bar{\beta}_p^{2\omega} = \beta_p^{2\omega} f_{i2} f_{j1} f_{k1},$$

where the $f_{\alpha\beta}$ are local field factors* (α = x, y, z; β = 1, 2) [122] and the direction cosines (the cosines of the angles between the direction $R_q R_t$ and the axes X, Y, Z of the crystal coordinate system) are l_p, m_p, n_p. Then, e.g., the z-component of the second-order dipole moment induced by the field $E = \{E_x, E_y, E_z\}$ for the p-th oscillator is equal to

$$(\Delta\mu_p^{2\omega})_z = [\bar{\beta}_p^{2\omega}(E_x l_p + E_y m_p + E_z n_p)^2]n_p = \bar{\beta}_p^{2\omega} n_p l_p^2 E_x^2 + \bar{\beta}_p^{2\omega} n_p m_p^2 E_y^2 + \bar{\beta}_p^{2\omega} n_p^3 E_z^2 + \ldots$$

and the z-component of the quadratic polarizability of the crystal is

$$P_z^{2\omega} = \chi_{zxx}^{2\omega} E_x^2 + \chi_{zyy}^{2\omega} E_y^2 + \chi_{zzz}^{2\omega} E_z^2 + \ldots,$$

---

*In the general case the $f_{\alpha\beta}$ are determined by the so-called Lorentz structure (lattice) sums and factors [146, 147].

where

$$\chi_{zxx}^{2\omega} = V_0^{-1} \sum_{p=1}^{x} \tilde{\beta}_p^{2\omega} n_p l_p^2, \qquad \chi_{zyy}^{2\omega} = V_0^{-1} \sum_{p=1}^{x} \tilde{\beta}_p^{2\omega} n_p m_p^2$$

$$\chi_{zzz}^{2\omega} = V_0^{-1} \sum_{p=1}^{x} \tilde{\beta}_p^{2\omega} n_p^3. \tag{31}$$

It is easy to see that in this notation the tensor $\chi^{2\omega}$ vanishes for a centrosymmetric distribution of the oscillators and is always symmetric in its three indices, although it is dispersive because of the frequency dependence of $\tilde{\beta}^{2\omega}$.

In spite of the strong dependence of the results obtained by applying the oscillator model on the choice of the structural elements and orientations of the corresponding oscillators of greatest importance in second-harmonic generation, as well as on the local field factors, and in spite also of the assumption that $\tilde{\beta}_{ijk}^{2\omega} \neq 0$ only for $i = j = k$ ($\mathbf{i} \parallel \mathbf{R_q R_t}$), this model can be useful in a large number of cases. Whereas the restrictions imposed on the tensor $\chi_{ijk}^{2\omega}$ by the point symmetry of the medium make it possible only to see which components are nonzero, the oscillator model often makes it possible to establish relations among the components and take into account the effect on them of the crystal structure.

We consider the crystals of the benzene derivatives investigated by us in Chap. 3 as well as in [43], viz., m-dihydroxybenzene, m-toluenediamine, m-dinitrobenzene, m-nitroaniline, 2-chloro-4-nitroaniline, and 2-bromo-4-nitroaniline. They all have the point group symmetry mm2. Four molecules are present in a unit cell in which the interaction of the π-electrons of the substituents in the meta position* is relatively weak. Making use of this fact and the results of the previous section, we express $\beta_V^{2\omega}$ in m-dihydroxybenzene, m-toluenediamine, m-dinitrobenzene, and m-nitroaniline as the sum of the $\beta_V^{2\omega}$ of the pairs of fragments $\phi - R_i$ ($R_i = $ OH, NH_2, NO_2), while we set $\beta_V^{2\omega}$ for 2-chloro- and 2-bromo-4-nitroaniline equal to the value of $\beta_V^{2\omega}$ for p-nitroaniline, $H_2N - \phi - NO_2$. Knowing the direction cosines n_i of the bonds joining the fragments and the local field factors $f_{\alpha\beta}$, it is possible using $\beta_V^{2\omega} \simeq \beta_{333}^{2\omega}$ (cf. Table 17) to find as in [138] the vector parts of the susceptibilities $\chi_V^{2\omega}$ of all six crystals. On the other hand, if we assume the effective oscillators $\phi - R_i$ and $H_2N - \phi - NO_2$ to be completely anisotropic ($\beta_P^{2\omega} \simeq \beta_{P333}^{2\omega}$), the components l_i, m_i, n_i, $f_{\alpha\beta}$, V_0 can be found using $\chi_{ijk}^{2\omega}$ and the data of Table 17. Finally, if at least one of the values l_i, m_i, n_i, $f_{\alpha\beta}$, is known, the $\beta^{2\omega}$ can be found from $\chi_{ijk}^{2\omega}$ and V_0 for the same oscillators.

Using the oscillator model for a molecular crystal, we calculated the ratio of all the nonzero independent tensor components of the quadratic optical susceptibility $\chi_{zxx}^{2\omega} : \chi_{zyy}^{2\omega} : \chi_{zzz}^{2\omega} \equiv$ F† of the six benzene derivatives given above, and found $|\chi_{zzz}^{2\omega}|$ from the known values $|\beta_p^{2\omega}|$. It was assumed for m-nitroaniline, in accordance with Table 17, that $|\beta^{2\omega} (\phi - NO_2)| \simeq 2|\beta^{2\omega} (\phi - NH_2)|$. Since we did not know the local field factors, we ignored their anistropy in determining F, and when estimating $|\beta_P^{2\omega}|$ we used the simplest form $(N_{\alpha\beta}^2 + 2)/3$ for the $f_{\alpha\beta}$.

Table 18 shows the direction cosines of the effective oscillators of the "basis" molecules[56] of the compounds investigated, the values of F found from them, the experimental values of $|\chi_{ijk}^{2\omega}|$ in units of $\chi_{zxy}^{2\omega}$ (for KDP), and also the values for $|\beta_p^{2\omega}|$ obtained. At the end

---

*That the $\phi - R_i$ systems in the meta-disubstituted benzenes are quasi-independent is indicated by the additivity of the group dipole moments and the IR and UV spectra.

†F in general depends on the frequency of the main radiation and its second harmonic.

TABLE 18

Crystal	l_1	l_2	m_1	m_2	n_1	Literature
m-Dihydroxybenzene	+0.850	−0.351	−0.474	+0.711	−0.287	[43. 138. 148]
m-Toluenediamine	+0,266	−0.517	−0,938	+0,534	+0,224	[90. 97]
m-Dinitrobenzene	−0.992	+0.419	−0,123	+0,854	−0,041	[43. 88. 96]
m-Nitroaniline	+0.798	−0.677	+0,580	+0.210	+0.163	[43. 138. 83]
2-Chloro-4-nitro-aniline	+0.580	—	−0,758	—	+0.282	[43. 149]
2-Bromo-4-nitro-aniline	+0.580	—	−0,758	—	+0.282	[43]

Crystal	n_2	V_0, Å³	F	$\lvert \chi^{2\omega}_{zxx} \rvert$	Literature
m-Dihydroxybenzene	−0,607	568,0	1.1 : 1.5 : 1.0	1.3±0.4 1,4±0.3	[43. 138. 148]
m-Toluenediamine	+0,669	645,5	0.6 : 1,3 : 1.0	0.9±0.2	[90. 97]
m-Dinitrobenzene	−0,323	708,8	2.8 : 7.0 : 1.0	<2.5 3.0±0.5	[43. 88. 96]
m-Nitroaniline	−0.705	638,6	1.1 : 0.1 : 1,0	51.0±7.6 21±4	[43. 138. 83]
2-Chloro-4-nitro-aniline	—	733,6	4,2 : 7.2 : 1,0	15.0±2.2	[43. 149]
2-Bromo-4-nitro-aniline	—	733,6	4,2 : 7.2 : 1,0	15,0±4.5	[43]

Crystal	$\lvert \chi^{2\omega}_{zyy} \rvert$	$\lvert \chi^{2\omega}_{zzz} \rvert$	$\lvert \beta^{2\omega}_p \rvert \cdot 10^{31}$, cgse·cm³	Literature
m-Dihydroxybenzene	1.6±0,5 3.5±0.7	1.7±0,5 2,8±0,6	1.6	[43. 138. 148]
m-Toluenediamine	2.6±0.5	1.8±0.4	2,2	[90. 97]
m-Dinitrobenzene	4.5±2.2 6.2±1.1	1.6±0.8 1,7±0,4	34	[43. 88. 96]
m-Nitroaniline	<1	23.0±3.5	15; 30	[43. 138. 83]
2-Chloro-4-nitro-aniline	2.3±0,5 35.0±5,2	25±5 6.5±1.0	200	[43. 149]
2-Bromo-4-nitro-aniline	35.0±10,5	6,8±2.0	210	[43]

of the table references are given for each compound to the papers in which $\lvert \chi^{2\omega}_{ijk} \rvert$ was measured and the crystal structure determined, the structural reference being given last. The fact that the 2-chloro- and 2-bromo-4-nitroaniline crystals have the same structure was used in the calculations.

The values of F found reflect quite well the true anisotropy of $\chi^{2\omega}$, and the values of $\lvert \beta^{2\omega}_p \rvert$ are in perfectly satisfactory agreement with the corresponding values in Table 17. The differences between $\beta^{2\omega}$ ($\varphi - NH_2$) for m-toluenediamine ($2.2 \cdot 10^{-31} \cdot cm^3$ cgse) and m-nitroaniline ($15 \cdot 10^{-31} cm^3$ cgse) are probably due to the roughness of our model. However, these differences might also be caused by a larger interaction of the amino group with the ring (and, to a lesser extent, with the nitro group) in m-nitroaniline than in m-toluenediamine.

The oscillator model also enables us to find $\chi^{2\omega}_{ijk}$ for crystals of p-nitro-p'-methylbenzalaniline.

As was already noted, the most efficient second-harmonic generation of light from a neodymium laser [$I_2 \simeq 300 I_2$(KDP)] among the organic compounds is observed in powders of p-nitro-p'-methylbenzalaniline, although it has so far not been possible to grow sufficiently large single crystals from this substance. Recent investigations [150] have shown that crystals of p-nitro-p'-methylbenzalaniline possess two modifications, of which only the monoclinic, acentric modification is responsible for second-harmonic generation. In this modification, the molecules of p-nitro-p'-methylbenzalaniline have an almost planar structure [151] which favors the formation of an extended system of conjugated bonds between a weak electron donor (CH_3) and a strong electron acceptor (NO_2) group. One might expect that the quadratic polarizability of molecules of p-nitro-p'-methylbenzalaniline, as in molecules of p-nitro-p'-dimethylaminostilbene [152] having a similar geometry, is markedly anisotropic and that $\beta^{2\omega}$ is in fact determined by the component $\beta_{333}^{2\omega}$ alone (the 3-axis lies along the long axis of the molecule), the value of which has the order of magnitude 10^{-29} cgse $\cdot$ cm^3.

The crystal symmetry m of p-nitro-p'-methylbenzalaniline admits the existence of six independent components of $\chi^{2\omega}$: $\chi_{xxx}^{2\omega}$, $\chi_{xyy}^{2\omega}$, $\chi_{xzz}^{2\omega}$, $\chi_{yxx}^{2\omega}$, $\chi_{yyy}^{2\omega}$, and $\chi_{yzz}^{2\omega}$ in the region of transparency. Using expression (31), the $\chi_{ijj}^{2\omega}$ (i = x, y; j = x, y, z) can be written as

$$\chi_{ijj}^{2\omega} = f_{i2}(f_{j1})^2 V_0^{-1} \sum_{p=1}^{Z} \beta_p^{2\omega} n_{pi} n_{pj}^2. \tag{32}$$

Putting $f_{i2} \simeq f_{j1} \simeq (N_{av}^2 + 2)/3$, $N_{av} \simeq 1.6$, $\beta_p^{2\omega} \simeq 10^{-29}$ cgse $\cdot$ cm^3, $V_0 = 6.4 \cdot 10^{-22}$ cm^3 and substituting into (32) the values $n_{px} = 0.753$, $n_{py} = 0.642$, $n_{pz} = \pm 0.234$ found from the atomic coordinates [151], we obtain

$$\chi_{xxx}^{2\omega} : \chi_{xyy}^{2\omega} : \chi_{xzz}^{2\omega} = \chi_{yxx}^{2\omega} : \chi_{yyy}^{2\omega} : \chi_{yzz}^{2\omega} \simeq n_{px}^2 : n_{py}^2 : n_{pz}^2 \simeq 10:8:1,$$

and $\chi_{xxx}^{2\omega} \simeq \chi_{yxx}^{2\omega} \simeq 40 \cdot 10^{-9}$ cgse.

Still larger optical nonlinearities and extremely efficient generation of the second harmonic of laser light should be expected for crystals of p-nitro-p'-dimethylaminobenzalaniline, $O_2NC_6H_4CH = NC_6H_4N(CH_3)_2$. Indeed, these molecules, like p-nitro-p'-methylbenzalaniline molecules, have an almost planar structure [153] facilitating maximal conjugation of strong electron donor $[N(CH_3)_2]$ and electron acceptor (NO_2) groups, and has a markedly polar structure ($\mu \simeq 7$ D [154]). The structurally similar molecule $O_2NC_6H_4CH = CHC_6H_4N(CH_3)_2$ of p-nitro-p'-dimethylaminostilbene with $\mu \simeq 7$ D [154] has $\beta^{2\omega} \simeq 10^{-28}$ cgse $\cdot$ cm^3 [152]* taking into account that $\beta^0 = 2\beta^{2\omega}$. However, in spite of the very high nonlinear molecular polarizability, only relatively weak generation of the second harmonic of neodymium laser light (cf. Table 4) is observed in powders of p-nitro-p'-dimethylaminobenzalaniline. For this reason it is possible to suggest that the position of the molecules in the unit cell is unfavorable for $\chi^{2\omega}$ ($n_p \ll 1$).

Thus, although a large number of assumptions lie at the basis of the oscillator model, this model nonetheless makes it possible to establish the role played by crystal structure in the formation of optical nonlinearities and reach definite conclusions concerning the tensor components of the quadratic susceptibility, without having to measure them.

§4. Second-Harmonic Generation and Some Properties of the Molecular and Crystal Structure of Organic Compounds

Correlating the quadratic susceptibility of crystals with their structure is of great importance for studying the nature of the optical nonlinearity of substances and systematically seeking out new materials for efficient conversion of the frequency of laser light.

---

*In crystals of p-nitro-p'-dimethylaminostilbene, second-harmonic generation is not observed, apparently because of a centrosymmetric structure.

In the preceding sections of this chapter we attempted to find general relations between the nonlinear properties of molecules and the character of their electron orbitals, and we showed using the oscillator model how the crystal structure of the substance influences the quadratic susceptibility. Allowance for the contributions to different molecular states of the electron configurations with charge transfer enabled us to relate the quadratic susceptibility of organic crystals in the general case with the presence of heteroatoms in the molecules and to explain the large nonlinearities of a number of compounds with highly asymmetric π-electron distribution in terms of acceptor–donor interaction of groups of atoms along chains of conjugated bonds. It was demonstrated that the orientation of the molecules (effective oscillators) relative to the crystallographic axes plays a role in the formation of quadratic susceptibility at least as important as that played by asymmetry and extendedness of the electron orbitals.

One of the main factors influencing the linear and nonlinear polarizability of organic molecules in the visible and near UV regions of the spectrum is conjugation caused by delocalization (mobility) of the π-electrons. Since π-conjugation shows up in the spectra, the dipole moments, and other characteristics of the molecular electron structure, the polarizabilities must also be closely related to these characteristics. Let us consider, e.g., some aspects of this problem as they relate to spectra for Raman scattering of light (RS).

During π-conjugation of molecular fragments, the π-clouds become combined into a single π-system. Here both the polarizability itself (the dimensions of the system increase) and its anisotropy (the dimensions of the system increase only along the conjugated chain) increase. The increased anisotropy leads on the other hand to a greater degree of depolarization and higher intensity of the RS lines. Table 19 compares the values $\beta^{2\omega} \approx \beta^{2\omega}_{333}$ already cited by us for some compounds and the intensity coefficients of the RS line of the benzene ring in the region 1570–1610 cm^{-1} (I^S_{1600}) from [155] and the RS lines of the completely symmetric vibrations of the nitro group ($I^S_{NO_2}$) from [156]. As was to be expected, the parameters $\beta^{2\omega}$ and I^S vary with the same slope.

It was shown in [157] that the degree of depolarization ρ_{NO_2} of the RS line for a completely symmetric stretching vibration of the nitro group increases with increasing conjugation of this group with the π-electron system and with increasing polarizability of the whole molecule. To the limiting value $\rho_{NO_2} = 0.5$ in molecules like p-nitrodimethylaniline, p-nitroaniline, etc., in which the polarizability and its coordinate derivative $\partial\alpha/\partial q$ are highly anisotropic, we can associate the completely anisotropic ellipsoid $\partial\alpha/\partial q$ of the nitro group which lies along the axis of symmetry of the NO_2 group (compare with the oscillator model in Sec. 3). In the meta derivatives of nitrobenzene, where the groups are weakly conjugated, ρ_{NO_2} does not depend on

TABLE 19

Structural formula	$\beta^{2\omega} \cdot 10^{31}$, cgse $\cdot$ cm^3	I^S_{1600}	$I^S_{NO_2}$	Structural formula	$\beta^{2\omega} \cdot 10^{31}$, cgse $\cdot$ cm^3	I^S_{1600}	$I^S_{NO_2}$
⬡Cl	1.8	35	—	⬡COCH$_3$	13	190	—
⬡Br	2	37	—	⬡NO$_2$	20	200	$7 \cdot 10^2$
⬡OH	2.2	38	—	O$_2$N⬡CH=N⬡CH$_3$	~100	—	$1.8 \cdot 10^4$
⬡OCH$_3$	4	50	—	O$_2$N⬡NH$_2$	200	—	$2 \cdot 10^4$
⬡NH$_2$	8.1	100	—	O$_2$N⬡CH=CH⬡N(CH$_3$)$_2$	~1000	—	$5 \cdot 10^6$

the electron donor character of the second substituent and is much smaller [158]. In accordance with this, the meta isomer of nitroaniline should have the smallest $\beta^{2\omega}$.

The dipole moments [159] can give a wealth of information concerning conjugation, eletron density distribution in the molecule, and conformation of a complex multiatomic system. A study of the bond lengths and valence angles of the molecules in a crystal also permits us to reach definite conclusions concerning the interactions in a conjugated system [160] and the contribution of one or another group of atoms to the nonlinear polarizability. Thus, a comparison of the bond lengths for $C-NH_2$ in m-toluenediamine (1.41 Å) and m-nitroaniline (1.39 Å) indicates that there is more conjugation of the amino group with the benzene ring in the second compound. This bond is even shorter (1.36 Å) in p-nitrodimethylaniline and the interaction of the dimethylamino group with the ring is still stronger.

Special features of the structure are often of help in understanding and correctly interpreting the results of second-harmonic generation in crystals. For example, data on crystal and molecular structure [161] indicate that the orientation about the Z axis of the two oscillators $(CH_3)_2N-\phi-NO_2$ in p-nitrodimethylaniline and $\phi-NO_2$ in nitromesitylene is unfavorable for generation of the second optical harmonic; the data also indicate that the nitro groups lie outside the plane of the aromatic core in nitromesitylene and 1,8-dinitronaphthalene, which decreases $\beta^{2\omega}$. Indeed, because of the point symmetry of crystals of p-nitrodimethylaniline (2) and nitromesitylene (mm2), all their $\chi_{ijk}^{2\omega}$ in the notation (31) contain $n_p \ll 1^*$ and the fact that the π-orbitals of the nitro group and benzene ring in nitromesitylene and 1,8-dinitronaphthalene are noncoplanar leads to a smaller nonlinear polarizability of these molecules, due to reduced mixing of the molecular electron configurations. The efficiency of second-harmonic generation in the powders of the three above compounds (cf. Tables 3 and 5) accords with our conclusion that the values of $\chi^{2\omega}$ for these substances are small owing to special features of the crystal and molecular structure.

A feature of organic compounds of importance for generation of the second optical harmonic is the fact that, as already emphasized, they are approximately twice as likely as inorganic compounds to form noncentrosymmetric structures, even though, of the molecular crystals, only $\sim 25\%$ are acentric, having the symmetries $P2_12_12_1$, $P2_1$, $Pna2_1$, etc. [64]. It is known [4] that a molecule having an inversion center retains it in a crystal field. Therefore, noncentrosymmetric molecules must be chosen to investigate the generation of the second harmonic of laser light. Increased proper asymmetry of the molecules reduces the probability that they will form crystals with an inversion center. Crystals of enantiomeric compounds with an asymmetric carbon molecule crystallize without an inversion center. The space groups $P2_1$ and $P2_12_12_1$ are the most probable ones for such crystals.

We noted an interesting "meta effect" in [63] for disubstituted benzenes with the general formula XC_6H_4Y. Namely, most of the ortho and para derivatives crystallize in the centrosymmetric classes, whereas the meta–disubstituted benzenes, which because of hydrogen bond do not form dimers with an inversion center, crystallize in the rhombopyramidal class mm2 [68] and give rise to frequency doubling of laser light. For example, all the crystals of the meta isomers given in Table 20 of the disubstituted benzenes XC_6H_4Y, except for nitrobenzoic acid ($X = NO_2$, $Y = COOH$) which forms centrosymmetric dimers, have point group symmetry mm2, and their powders generate the second harmonic of light from a neodymium laser with relative intensity I_2.

---

*A substantially more favorable position of oscillators with smaller $\beta^{2\omega}$, which we do not take into account, can give rise to a contribution to $\chi^{2\omega}$ which is comparable to or even greater than the contribution from oscillators with large $\beta^{2\omega}$.

TABLE 20

Substituent		Isomers					
		ortho-		meta-		para-	
X	Y	class	I_2, rel. units	class	I_2, rel. units	class	I_2, rel. units
NO_2	NH_2	$2/m$	0	$mm2$	42	$2/m$	0
NO_2	COH	2	17	$mm2$	32	$mm2$	20
NO_2	NO_2	$2/m$	0	$mm2$	28	$2/m$	0
NO_2	COOH	$\bar{1}$	0	$2/m$	0	$2/m$	0
NH_2	OH	mmm	0	$mm2$	2	$mm2$	0.3
OH	OH	$2/m$	0	$mm2$	1.5	$\bar{3}$	0
NH_2	NH_2	$2/m$	0	$mm2$	0.2	$2/m$	0

Although the mechanisms for second-harmonic generation and the piezoeffect are different, the relation between the induced polar vector P and the quantities determining the response of an anisotropic medium is of the same character in either case. Both phenomena are described by a tensor of the same type and are associated with asymmetric crystal structures. Therefore, many of the principles used in seeking noncentrosymmetric piezoactive materials can also be used in choosing substances suitable for second-harmonic generation.

The correlation of dipoles localized in a crystal with the piezoeffect [162] and of quadratic optical nonlinearities with asymmetric electron density distribution in the ground and the excited states of a system suggests that the ten polar classes 1, 2, 3, 4, 6, m, mm2, 3m, 4mm, 6mm favorable for the piezoelectric effect [163] will also be favorable for frequency doubling of laser light.

Numerous meta-disubstituted benzenes in which second-harmonic generation is observed, e.g., m-nitroaniline, m-dinitrobenzene, m-phenylenediamine, m-dihydroxybenzene, m-aminophenol, m-nitrophenol, are members of a genetic series of piezoelectrics [13]. Homological isomorphism can serve as the foundation of the genetic principle for seeking crystals without an inversion center and which are efficient in frequency doubling; homological isomorphism should also play an important role in the study of second-harmonic generation in mixed crystals.

The structural features of molecular crystals should be taken into account both when looking for acentric structures suitable for second-harmonic generation, and when studying such substances. In particular, if the crystal belongs to one of the point group symmetries (222, 422, 32, $\bar{6}$, 622, or $\bar{6}2m$) it is possible to say immediately that there is no noncritical phase synchronism in the crystal, and the presence in the crystal of pronounced cleavages may cause considerable ordering of microcrystals in samples and distort results of the estimation of $\chi^{2\omega}$ by the powder method. Bearing in mind the latter situation, we were able to predict correctly the existence of a direction of synchronism in crystals of m-toluenediamine which is nearly equal to the normal direction to one of the cleavage planes.

§5. More Precise Determination of the Symmetry

of Molecular Crystals Using Second-Harmonic

Generation of Laser Light

One of the first problems in x-ray structure studies of crystals, which is solved before the x-ray photographs are indexed, is to establish the Laue symmetry. As is well known [164], there are altogether eleven possible Laue classes. The point group symmetry of a crystal cannot be established from the Laue class. (It is not possible, e.g., to distinguish among the symmetries 2, m, or 2/m that a crystal belonging to the monoclinic crystal system might have, since the addition of an inversion center to the groups 2 or m transforms them into 2/m. In all three cases the Laue symmetry is the same.) To this end one studies the extinction of the x-ray reflections. However, not even the extinction pattern permits the determination of mirror reflection planes, rotation axes, and inversion centers. Thus, the space groups of symmetry Pa and $P2_1$ are indistinguishable from the centrosymmetric groups $P2/a$ and $P2_1/m$, respectively.

In order to ascertain the true symmetry of crystals with more precision, one studies the statistics of the intensity pattern of the diffracted spots, the external form of the crystal, etch figures, piezo- or pyroelectric properties, and rotation of the plane of polarization of light; radiofrequency methods are also used. Each of these methods has limitations associated with its specific features. For example, the symmetry of the external form may turn out to be too high, and the piezoelectric properties may be either small* due to proper electric conductivity, hygroscopicity, or small dimensions ($\hat{r} < \lambda_B/2$), or else masked by a secondary piezoelectric effect [165]. But even the presence of the piezoelectric effect only says that the crystal is acentric. If the crystal displays a pyroeffect, this indicates that there is no inversion center, plane of symmetry, or fourth-order symmetry axis perpendicular to the direction of the effect, but it says nothing about longitudinal planes of symmetry. Rotation of the plane of polarization of light only indicates the absence of such longitudinal planes and does not give information concerning the remaining symmetry elements.

Second-harmonic generation gives another method for accurately finding the symmetry of crystals. Indeed, the symmetry of the medium and the symmetry of the tensor $\chi_{ijk}^{2\omega}$ in the last two indices impose restrictions on the components $\chi_{ijk}^{2\omega}$ which are closely connected with the point symmetry of the crystal. Second-harmonic generation is forbidden in the case of a structure with an inversion center, and also when the crystal belongs to the class 432 or (when there is weak dispersion) to the classes 422 or 622. If the j-axis is a fourth-order symmetry axis, second-harmonic generation is not observed for a laser beam propagating along the j-axis. For nonpolar directions j, second-harmonic generation in the crystal due to the component $\chi_{jjj}^{2\omega}$ is also absent. In pyroelectric crystals, second-harmonic generation occurs when the laser light is polarized along any of the X, Y, or Z axes. Table 21 in [78] gives all the possible variants of second-harmonic generation in powders and single crystals which can be expected with allowance for dispersion for each Laue and crystal class when the direction of propagation s_1 of the laser beam (for single crystals) and the direction e_1 of the beam are parallel to the crystallographic Z axis.† Second-harmonic generation (SHG) which is permitted by a symmetry is indicated with a plus, and a minus sign means it is forbidden.

Modern methods for recording light signals make it possible to detect second-harmonic generation even when the nonlinearities are small. Therefore, if the results of observation of the second harmonic in single crystals for $s_1 \parallel Z$ and their powders are known (cf. Sec. 5, Chap. 2), it is in most cases possible to choose the true point group symmetry in a given Laue class using Table 21. Thus, the crystals of m-toluenediamine described in Chap. 3 were assigned the Laue class from Laue photographs. The presence of second-harmonic generation of light from a neodymium laser in the powder of this substance immediately ruled out the

---

*The Tul'skii spectrometer [166] is too sensitive for determining the precise lattice symmetry since it registers the local piezoelectric ordering.

†In the tetragonal, trigonal, and hexagonal system, the principal symmetry axis Z coincides with the optical axis.

TABLE 21

Laue class	$C_i=\bar{1}$		$C_{2h}=2/m$			$D_{2h}=mmm$			$C_{4h}=4/m$			$D_{4h}=4/mmm$				$C_{3i}=\bar{3}$	
Symmetry point group	1	$\bar{1}$	2	m	$2/m$	222	$mm2$	mmm	4	$\bar{4}$	$4/m$	422	$4mm$	$\bar{4}2m$	$4/mmm$	3	$\bar{3}$
SHG in powder	+	−	+	+	−	+	+	−	+	+	−	+	+	+	−	+	−
SHG in single crystal, $s_1\|Z$	+	−	+	+	−	−	+	−	+	+	−	−	+	+	−	+	−
SHG in single crystal, $e_1\|Z$	+	−	+	+	−	−	+	−	+	−	−	−	+	−	−	+	−

Laue class	$D_{3d}=\bar{3}m$			$C_{6h}=6/m$			$D_{6h}=6/mmm$				$T_h=m3$		$O_h=m3m$		
Symmetry point group	32	$3m$	$\bar{3}m$	6	$\bar{6}$	$6/m$	622	$6mm$	$\bar{6}2m$	$6/mmm$	23	$m3$	432	$\bar{4}3m$	$m3m$
SHG in powder	+	+	−	+	+	−	+	+	+	−	+	−	−	+	−
SHG in single crystal, $s_1\|Z$	+	+	−	+	+	−	−	+	+	−	−	−	−	−	−
SHG in single crystal, $e_1\|Z$	−	+	−	+	−	−	−	+	−	−	−	−	−	−	−

centrosymmetric point group mmm, and second-harmonic generation when the laser beam was polarized along one of the proposed second-order axes enables us to assign the group mm2 as the point group symmetry of these crystals. The correctness of this symmetry group was confirmed in [97] by indexing the x-ray reflections, which revealed systematic extinctions in the reflections h0l (h = 2n + 1) and 0kl (k + l = 2n + 1) and the absence of extinction conditions among the reflections of general type and type hk0; the theory of dense packing of molecules was used here [4].

CONCLUSION

The main results of this paper can be formulated as follows:

1. Second-harmonic generation of light from neodymium and ruby lasers was investigated in the powders of 300 organic compounds of various classes. Generation was discovered in the powders of 80 compounds. The crystals of the remaining substances were evidently centrosymmetric. For most of the substances, a drop in the efficiency of frequency doubling caused by absorption at the frequency of the harmonic was observed when light from a neodymium laser was replaced by light from a ruby laser. Ten organic compounds were found to be of promise in doubling the frequency of high-power laser light with $\lambda \sim 1 \ \mu$m.

2. The possibility was studied experimentally of using observations of second-harmonic generation with $\lambda = 0.53 \ \mu$m in powders to determine noncentrosymmetry of crystals. A comparison of the results obtained for a large number of organic substances with data from x-ray and piezoelectric studies showed that the laser method has a number of advantages (it is fast, highly sensitive, and simple). It was shown that it is almost always possible to choose the true point group symmetry from a given Laue class from information obtained by observing second-harmonic generation of polarized light in single crystals and powders.

3. The absorption spectra of polarized light, dispersion of the main indices of refraction, and the relative values of the moduli of all components of the quadratic susceptibility tensor $\chi^{2\omega}$ ($\lambda_1 = 1.06 \ \mu$m) were found for crystals of m-dinitrobenzene and m-toluenediamine. The quantities $\chi^{2\omega}_{ijk}$ for crystals of m-dinitrobenzene are in good agreement with independent measurements of other workers. The dispersion and birefringence of crystals of m-dinitrobenzene and m-toluenediamine were used to calculate all possible directions of phase synchronism at $\lambda = 1.06 \ \mu$m, as well as the parameters α_m, $\delta\alpha_m$, $\delta\lambda_1$, etc., characterizing collinear phase synchronism for interactions of type I in the principal planes of the crystals. The experimental values for α_m and $\delta\alpha_m$ correspond to the results of calculations. The possibility was demonstrated of using crystals of m-toluenediamine to double the frequency of high-power laser radiation with $\lambda = 1.06 \ \mu$m.

4. Equations were obtained which determine all possible directions of the wave normals of the principal and transformed light for second-harmonic generation in biaxial crystals in the presence of noncollinear phase synchronism. The calculations for crystals of m-dinitrobenzene and m-toluenediamine agree with the experimental results.

5. It was shown that the quadratic susceptibility of organic crystals in the general case is related to the presence of heteroatoms in the molecules of the crystal and is caused both by asymmetry and extendedness of the electron orbitals and by the orientation of the molecules relative to the crystallographic axes.

6. A qualitative estimate of the optical nonlinearity of some monosubstituted benzenes was given taking into account the role of configurations with charge transfer. It was concluded that the quadratic polarizability increases in the series of compounds with (σ, π), (σ, π, l), (σ, π, n), and (σ, π, l, n) electron orbitals.

7. Using the oscillator model, it was shown for six molecular crystals that it is possible, given a known crystal structure, to find relations among the components of the quadratic susceptibility tensor and also among the nonlinear polarizabilities of the effective oscillators.

Our investigations show that molecular crystals of organic compounds may find extensive application in nonlinear optics, provided techniques are further improved for their growth, handling, and protection from effects of exposure to the atmosphere.

I express my sincere thanks to my advisors A. M. Prokhorov and G. P. Shipulo for their constant interest in this work and their support, A. P. Skoldinov and E. A. Smirnov for providing a large number of organic substances for the investigations, and also V. A. Chikov for assistance with this work.

LITERATURE CITED

1. S. A. Akhmanov and R. V. Khokhlov, Problems in Nonlinear Optics [in Russian], VINITI, Moscow (1964).
2. P. A. Franken, A. E. Hill, C. W. Peters, and G. Weinreich, Phys. Rev. Lett., 7:118 (1961).
3. V. S. Suvorov and A. S. Sonin, Kristallografiya, 11:832 (1966).
4. A. I. Kitaigorodskii, Molecular Crystals [in Russian], Nauka, Moscow (1971).
5. G. V. Bokii, Crystal Chemistry [in Russian], Izd. Mosk. Gos. Univ. (1960).
6. A Short Chemical Encyclopedia [in Russian], Sov. Éntsikl., Moscow (1964).
7. Y. Okaya and R. Pepinsky, Acta. Crystallogr., 10:324 (1957).
8. W. Nowacki and H. Jaggi, Z. Kristallogr., 109:272 (1957).
9. T. Penkalya, Outlines of Crystal Chemistry [in Russian], Khimiya, Leningrad (1974).
10. J. Donohue and A. Caron, Acta Crystallogr., 17:663 (1964).
11. K. Hirayama, Handbook of UV and Visible Absorption Spectra of Organic Compounds, Plenum Press, New York (1971).
12. A. N. Winchell, The Optical Properties of Organic Compounds, Academic Press, New York (1954).
13. I. S. Rez, Kristallografiya, 5:63 (1960).
14. W. Nowacki, Am. Cryst. Assoc. Monograph, Vol. 6 (1967).
15. R. Yu. Orlov, Candidate's thesis, Mosk. Gos. Univ. (1964).
16. K. E. Rieckhoff and W. L. Peticolas, Science, 147:610 (1965).
17. P. M. Rentzepis and Y. H. Pao, Appl. Phys. Lett., 5:156 (1964).
18. G. H. Heilmeier, N. Ockman, R. Braunstein, and D. A. Kramer, Appl. Phys. Lett., 5:229 (1964).
19. A. A. Filimonov, V. S. Suvorov, and I. S. Rez, Zh. Éksp. Teor. Fiz, 56:1519 (1969).
20. S. K. Kurtz and T. T. Perry, J. Appl. Phys., 39:3798 (1968).
21. A. Graja, Phys. Status Solidi, 27:K93 (1968); Acta. Phys. Pol., A37:539 (1970).
22. S. A. Akhmanov and A. S. Chrikin, Statistical Phenomena in Nonlinear Optics [in Russian], Izd. Mosk. Gos. Univ. (1971).
23. J. R. Gott, M. J. P. Musgrave, and J. N. Murrett, Mol. Phys., 12:295 (1967).
24. R. Yu. Orlov, Kristallografiya, 11:463 (1966).
25. A. A. Filimonov, V. S. Suvorov, R. I. Rez, et al., Proc. Fourth All-Union Symp. on Nonlinear Optics [in Russian], Kiev (1968).
26. M. Bass, D. Bua, R. Mozzi, and R. Monchamp, Appl. Phys. Lett., 15:393 (1969).
27. L. D. Derkacheva, A. I. Krymova, and N. P. Sonina, Pis'ma Zh. Éksp. Teor. Fiz., 11:469 (1970).
28. B. L. Davydov, L. D. Derkacheva, V. V. Dunina, et al., Pis'ma Zh. Éksp. Teor. Fiz., 12:24 12:24 (1970).

29. J. P. Hazan and J. Haisma, Opt. Commun., 2:343 (1970).
30. B. L. Davydov, L. D. Derkacheva, V. V. Dunina, et al., Opt. Spektrosk., 30:503 (1971).
31. J. Jerphagnon, IEEE J. Quant. Electron, QE-7:42 (1971).
32. E. A. Tikhonov and M. T. Shpak, Ukr. Fiz. Zh., 17:190 (1972).
33. B. L. Davydov, V. F. Zolin, L. G. Koreneva, et al., Zh. Prikl. Spektrosk., 17:413 (1972).
34. G. P. Bolognesi, S. Mezzetti, and F. Pandarese, Opt. Commun., 8:267 (1973).
35. B. L. Davydov, V. F. Xolin, L. G. Koreneva, et al., Zh. Prikl. Spektrosk., 20:516 (1974).
36. L. G. Koreneva, B. L. Davydov, M. E. Zhabotinskii, and V. F. Zolin, IRÉ Preprint No. 105, Moscow (1972).
37. B. L. Davydov, V. V. Dunina, V. F. Zolin, and L. G. Koreneva, Opt. Spektrosk., 34:267 (1973).
38. B. L. Davydov, V. V. Dunina, V. F. Zolin, et al., Opt. Spektrosk., 32:225 (1972).
39. B. L. Davydov, L. G. Koreneva, and M. A. Samokhina, IRE Preprint No. 104, Moscow (1972).
40. B. L. Davydov, V. F. Zolin, L. G. Koreneva, and M. A. Samokhina, Zh. Prikl. Spektrosk., 18:156 (1973).
41. K. Ito, Y. Kusuhara, K. Hamano, and S. Sawada, Jpn. J. Appl. Phys., 13:1299 (1974).
42. P. D. Southgate and D. S. Hall, J. App. Phys., 42:4480 (1971).
43. P. D. Southgate and D. S. Hall, J. Appl. Phys., 43:2765 (1972).
44. J. G. Bergman, G. R. Crane, B. F. Levine, and C. G. Bethea, Appl. Phys. Lett., 20:21 (1972).
45. S. Singh, W. A. Bonner, T. Kyle, et al., Opt. Commun., 5:131 (1972).
46. W. A. Norland, Ferroelectrics, 2:57 (1971).
47. B. F. Levine and C. G. Bethea, Appl. Phys. Lett., 20:272 (1972).
48. J. Jerphagnon, Appl. Phys. Lett., 16:298 (1970).
49. J. R. Gott, J. Phys., B4:116 (1971).
50. B. L. Davydov, L. G. Koreneva, and E. A. Lavrovskii, Radiotekh. Élektron., No. 6:1313 (1974).
51. P. D. Southgate and D. S. Hall, Appl. Phys. Lett., 18:456 (1971).
52. N. A. Vlasenko, B. L. Davydov, and L. G. Koreneva, Izv. Vyssh. Uchebn. Zaved., Radiofiz., 16:363 (1973).
53. M. V. Hobden, J. Appl. Phys., 38:4365 (1967).
54. R. Yu. Orlov, Izv. Vyssh. Uchebn. Zaved., Radiofiz. 12:1351 (1969).
55. B. L. Davydov, M. E. Zhabotinskii, and V. F. Zolin, Pis'ma Zh. Éksp. Teor. Fiz., 13:336 (1971).
56. N. L. Boling, M. J. Crisp, and G. Dube, Appl. Opt., 12:650 (1973).
57. F. N. H. Robinson, Bell System Tech. J., 46:913 (1967).
58. F. Lebon, Phys. Status Solidi, 41:297 (1970).
59. L. G. Koreneva, V. F. Zolin, and B. L. Davydov, Molecular Crystals in Nonlinear Optics, [in Russian], Nauka, Moscow (1975).
60. S. J. Syvin, J. E. Rauch, and J. C. Decius, J. Chem. Phys., 43:4083 (1965).
61. V. D. Shigorin and G. P. Shipulo, Kratk. Soobshch. Fiz., No. 5, p. 59 (1970).
62. V. D. Shigorin and G. P. Shipulo, Kratk. Soobshch. Fiz., No. 11, p. 31 (1970).
63. V. D. Shigorin and G. P. Shipulo, Kratk. Soobshch. Fiz., No. 10, p. 34 (1971).
64. V. K. Bel'skii and P. M. Zorkii, Kristallografiya, 15:704 (1970).
65. A. V. Voskanyan, D. N. Klyshko, and V. S. Tumanov, Zh. Éksp. Teor. Fiz., 45:1399 (1963).
66. G. L. Gurevich and Yu. G. Khronopulo, Zh. Éksp. Teor. Fiz., 51:1497 (1966).
67. V. D. Shigorin and G. P. Shipulo, Kvantovaya Élektron., No. 4, p. 116 (1972).
68. Landolt-Bornstein, New Series, Group III, Vol. 5, Parts a, b, Springer-Verlag, Berlin–Heidelberg–New York (1971).

69. A. Hettichand and H. Steinmetz, Z. Phys., 76:688 (1932).

70. V. A. Koptsik, K. A. Minaeva, A. A. Voronkov, et al., Vestn. Mosk. Gos. Univ., 6:91 (1958).

71. I. S. Rez, A. S. Sonin, E. E. Tsepelevich, and A. A. Filimonov, Kristallografiya, 4:65 (1959).

72. Shok Le Nguen and G. A. Gol'der, Zh. Strukt. Khim., 11:939 (1970).

73. L. A. Chetkina, O. A. Mukhno, and Z. I. Ezhkova, Zh. Strukt. Khim., 12:187 (1971).

74. H. Ranin, Radiat. Effects, 4:167 (1970).

75. T. P. Belikova and E. A. Sviridenkov, Pis'ma Zh. Éksp. Teor. Fiz., 5:29 (1967).

76. I. I. Konilenko, P. A. Korotkov, and V. I. Malyi, in: Quantum Electronics, No. 3 [in Russian], Naukova Dumka, Kiev (1969), p. 120.

77. V. D. Shigorin and G. P. Shipulo, Kristallografiya, 18:557 (1973).

78. V. D. Shigorin and G. P. Shipulo, Kristallografiya, 19:1006 (1974).

79. A. Coda, M. Fumagalli, F. Pandarese, and L. Ungaretti, Acta. Crystallogr., A31:S208 (1975).

80. V. E. Zavodnik, Z. P. Povet'eva, and Z. V. Zvonkova, Kristallografiya, 20:839 (1975).

81. D. Kalymnios, J. Phys., D5:667 (1972).

82. S. Ayers, M. M. Faktor, D. Marr, and J. L. Stevenson, J. Mater. Sci., 7:31 (1972).

83. J. L. Syeveson and A. C. Skapski, J. Phys., C5:L233 (1972).

84. J. L. Steveson, S. Ayers, and M. M. Faktor, J. Phys. Chem. Solids, 34:235 (1973).

85. J. L. Steveson, J. Phys., D6:L13 (1973).

86. G. S. Belikova, M. P. Golovei, V. D. Dhigorin, and G. P. Shipulo, in: Proc. Sixth All-Union Conf. on Nonlinear Optics [in Russian], Minsk (1972), p. 55.

87. G. S. Belikova, M. P. Golovei, V. D. Shigorin, and G. P. Shipulo, Krat. Soobshch. Fiz., No. 6, p. 24 (1973).

88. G. S. Belikova, M. P. Golovei, V. D. Shigorin, and G. P. Shipulo, Opt. Spektrosk., 38:779 (1975).

89. S. S. Grazhulene, L. A. Musikhin, V. Sh. Shekhtman, et al., in: Proc. Seventh All-Union Conf. on Coherent and Nonlinear Optics [in Russian], Moscow (1974), p. 30.

90. V. D. Shigorin, G. I. Shipulo, S. S. Grazhulene, et al., Kvantovaya Élektron., 2:2539 (1975).

91. E. M. Archer, Proc. Roy. Soc., A188:51 (1946).

92. K. Nakamoto, J. Am. Chem. Soc., 74:392 (1952).

93. J. Trotter, Acta Crystallogr., 14:244 (1961).

94. G. S. Belikova, L. E. Kraeva, and V. D. Shigorin, Inventor's Certificate No. 459956, Byull. Izobr., No. 5 (1975).

95. S. S. Grazhulene, L. A. Musikhin, G. F. Telegin, and V. D. Shigorin, Inventor's Certificate No. 496043, Byull. Izobr., No. 47 (1975).

96 J. Trotter and C. S. Williston, Acta Crystallogr., 21:285 (1966).

97. O. S. Filipenko, V. I. Ponomarev, V. D. Shigorin, et al., Kristallografiya, 20:937 (1975).

98. S. V. Grum-Grzhimailo, Instruments and Methods for Optical Investigation of Crystals [in Russian], Nauka, Moscow (1972).

99. I. S. Rez, Usp. Fiz. Nauk, 93:633 (1967).

100. A. V. Shubnikov, Fundamentals of Optical Crystallography [in Russian], Izd. Akad. Nauk. SSSR, Moscow (1958).

101. N. M. Melankholin, Methods for Investigating Optical Properties of Crystals [in Russian], Nauka, Moscow (1970).

102. L. S. Khrenov, Seven-Place Tables of the Trigonometric Functions [in Russian], Gos. Izd. Tekh. Teor. Lit., Moscow (1956).

103. W. L. Bond, J. Appl. Phys., 36:1674 (1965).

104. M. N. Datta, Indian J. Phys., 21:303 (1947).

105. M. A. Lasheen and A. M. Abdeen, Acta. Crystallogr., A28:245 (1972).
106. P. D. Maker, R. W. Terhune, M. Nisenoff, and C. M. Savage, Phys. Rev. Lett., 8:21 (1962).
107. J. Jerphagnon and S. K. Kurtz, J. Appl. Phys., 41:1667 (1970).
108. Y. Uematsu, Jpn. J. Appl. Phys., 12:1257 (1973).
109. G. E. Francois, Phys. Rev., 143:597 (1966).
110. A. Savage, J. Appl. Phys., 36:1496 (1965); G. D. Boyd, H. Kasper, and J. H. McFee, IEEE J. Quant. Electron., 7:563 (1971).
111. D. A. Kleinman, Phys. Rev., 126:1977 (1965).
112. R. C. Miller, Appl. Phys. Lett., 5:17 (1964).
113. V. D. Shigorin and G. P. Shipulo, Kvantovaya Élektron., 3:2048 (1976).
114. H. E. Bates, J. Opt. Soc. Am., 63:146 (1973).
115. B. V. Bokut', Zh. Prikl. Spektrosk., 7:621 (1967).
116. B. L. Davydov, V. F. Zolin, L. G. Koeneva, and E. A. Lavrovskii, Opt. Spektrosk., 39:713 (1975).
117. H. Ito, H. Naito, and H. Inaba, IEEE J. Quant. Electron., 10:247 (1974).
118. A. I. Kovrigin, N. K. Podsotskaya, and A. P. Sukhorukov, in: Nonlinear Optics [in Russian], Nauka, Novosibirsk (1968), p. 393.
119. G. D. Boyd and D. A. Kleinman, J. Appl. Phys., 39:3597 (1968).
120. G. S. Belikova, L. A. Kulevsky, Yu. N. Polivanov, et al., J. Raman Spectrosc., 2:493 (1974).
121. G. S. Belikova, S. S. Grazhulene, L. M. Doroshkin, et al., in: Proc. Eighth All-Union Conf. on Coherent and Nonlinear Optics, Vol. 1 [in Russian], Izd. "Metsniereba," Tbilisi (1976), p. 229.
122. J. Ducuing and C. Flytzanis, in: Optical Properties of Solids, F. Abeles, ed., Amsterdam (1972), p. 863.
123. V. D. Shigorin and G. P. Shipulo, in: Proc. Sixth All-Union Conf. on Nonlinear Optics [in Russian], Minsk (1972), p. 19.
124. V. D. Shigorin and G. P. Shipulo, Opt. Spektrosk., 34:151 (1973).
125. N. Bloembergen, Nonlinear Optics, Benjamin, New York (1965).
126. C. G. B. Garrett, IEEE J. Quant. Electron., 4:70 (1968).
127. K. Higasi, H. Baba, and A. Rembaum, Quantum Organic Chemistry, Interscience, New York (1965).
128. A. D. McLachlan, Mol. Phys., 2:271 (1959).
129. V. D. Shigorin and G. P. Shipulo, Zh. Prikl. Spektrosk., 20:720 (1974).
130. R. S. Mulliken, J. Chem. Phys., 7:20 (1939).
131. H. Ratajczak, J. Phys. Chem., 76:3000, 3991 (1972).
132. S. Nagakura, M. Kojima, and Y. Marugama, J. Mol. Spectrosc., 13:174 (1974).
133. H. Suzuki, Electronic Absorption Spectra and Geometry of Organic Molecules, Academic Press, New York–London (1967).
134. M. Godfrey and J. N. Murrell, Proc. Roy. Soc., A278:71 (1964).
135. G. V. Saidov and N. G. Bakhshiev, Dokl. Akad. Nauk SSSR, 175:1090 (1967).
136. V. I. Danilova, Author's Abstract of Doctoral Dissertation, Tomsk (1969).
137. S. Kobinata and S. Nagakura, J. Am. Chem. Soc., 88:3905 (1966).
138. J. L. Oudar and D. S. Chemla, Opt. Commun., 13:164 (1975).
139. J. Jerphagnon, Phys. Rev., B2:1091 (1970).
140. C. N. R. Rao, Ultraviolet and Visible Spectroscopy: Chemical Applications, Plenum Press, New York (1967).
141. K. B. Everard and L. E. Sutton, J. Chem. Soc., 2821 (1951).
142. L. Doub and J. M. Vandenbelt, J. Am. Chem. Soc., 69:2714 (1947).
143. B. F. Levine and C. G. Bethea, Appl. Phys. Lett., 24:445 (1974).

144. C. Flytzanis and J. Ducuing, Phys. Rev., 178:1218 (1969).
145. L. B. Meisner, Fiz. Tverd. Tela, 14:2220 (1972).
146. D. A. Dunmur, Mol. Phys., 23:109 (1972).
147. D. D. Karov, S. N. Koikov, and M. N. Rogova, Fiz. Tverd. Tela, 15:594 (1973).
148. G. E. Bacon and N. A. Curry, Proc. Roy. Soc., A235:552 (1956).
149. A. T. McPhail and G. A. Sim, J. Chem. Soc., 227 (1965).
150. V. I. Ponomarev, O. S. Filipenko, L. O. Atovmyan, et al., Kristallografiya, 22:394 (1977).
151. O. S. Filipenko, V. D. Shigorin, V. I. Ponomarev, et al., Kristallografiya, 22:534 (1977).
152. A. Schweig, Chem. Phys. Lett., 1:195 (1967).
153. P. Skrabal, J. Steiger, and H. Zollinger, Helv. Chim. Acta, 58:800 (1975).
154. Handbook of Dipole Moments [in Russian], "Vysshaya shkola," Moscow (1971).
155. P. P. Shorygin, Usp. Khim., 40:694 (1971).
156. P. P. Shorygin, V. P. Roshchupkin, V. A. Petukhov, and Z. S. Egorova, Zh. Fiz. Khim., 35:258 (1961).
157. Ya. S. Bobovich and M. V. Vol'kenshtein, Dokl. Akad. Nauk SSSR, 71:1045 (1950).
158. Ya. S. Bobovich and M. Ya. Tsenter, Opt. Spektrosk., 8:45 (1960).
159. V. I. Minkin, O. A. Osipov, and Yu. A. Zhdanov, Dipole Moments in Organic Chemistry, Plenum Press, New York (1970).
160. A. Domenicano, A. Vaciago, and C. A. Coulson, Acta Crystallogr., B31:221 (1975).
161. R. W. Wyckoff, in: Crystal Structures, Vol. 6, Parts 1, 2, Interscience, New York–London (1971).
162. V. A. Koptsik, Izv. Akad. Nauk SSSR, Ser. Fiz., 20:219 (1956).
163. I. S. Zheludev, Kristallografiya, 2:90 (1957).
164. G. V. Boikii and M. A. Porai-Koshits, A Practical Course in X-ray Analysis [in Russian], Izd. Mosk. Gos. Univ., Moscow (1951).
165. I. S. Rez, P'ezotekhnika, No. 1, p. 3 (1959).
166. B. V. Tul'skii, Zh. Strukt. Khim., 6:304 (1965).

CALCULATION BY THE WKB METHOD OF RESONATOR MODES FOR LASERS WITH AN ACTIVE MEDIUM

B. P. Kirsanov and A. M. Leontovich

The applicability of the classical WKB method is shown in the case of a complex potential (complex dielectric permittivity) for calculating the modes of a resonator with an active medium. It is demonstrated that the mode spectra and losses for a plane-parallel resonator with a quadratic or hypergeometric transverse distribution of the complex dielectric permittivity agree with the results of exact calculation. It is also shown that the modes of two-dimensional resonators with bounded mirrors which differ from plane-parallel can be calculated by the WKB method by reducing the problem to the case of a plane-parallel resonator containing a dielectric, the boundaries of the resonator being modelled by a suitably chosen jump of the imaginary part of the dielectric permittivity.

§1. Introduction

Although the theory of open resonators for frequencies in the optical range has in recent years undergone very rapid development (see, e.g., the surveys [1, 2]) in response to the vigorous development of quantum radiophysics, most of the work has been devoted to the theory of idealized models of empty resonators with no allowance made for their active medium. At the same time, in [3-5] it was shown that the field distribution and mode spectrum depend not only on the distribution inside the resonator of the real part of the dielectric permittivity ε', but also on its imaginary part ε''. This effect is seen most clearly in the mode loss spectrum [5]. The most general results on calculation of modes of resonators occupied by an inhomogeneous active medium are obtained by perturbation theory [4]. However, this method is quite cumbersome and does not always give results of a perspicuous form. In this connection it should be remarked that, in general, the use of precise methods for calculating the modes of ideal resonators in laser applications is not always effective, since under real conditions the active media (of solid lasers, for example) usually have inhomogeneities in ε, often of a random character, and it is difficult to take their effect into account in the precise theories. For this reason the development of simple, albeit approximate, methods of calculation of modes of resonators filled with a dielectric with an arbitrary distribution of ε is a problem of great current interest.

The Wentzel−Kramers−Brillouin (WKB) method (the "quasi-classical approximation") [6-8] is of great interest owing to its simple physical interpretation, the simplicity of its results, and its universality, which make it possible to estimate the effect of all sorts of perturbations. The modes of empty resonators which correspond to quantum mechanical problems with a purely real potential, have already been calculated by this method [9, 10].

The general theory of the use of the WKB method in the complex domain is not well developed (see, e.g., [11, 12]), but it is known that quantization (finding the frequency spectrum) is not possible for certain complex potentials [13]. Hence the application of the WKB method to specific complex situations requires special analysis.

In this paper it is shown that the WKB method can in principle be used to calculate the frequency spectrum, losses, and eigenfunctions of optical resonators given a sufficiently smooth distribution of a complex dielectric permittivity which changes sufficiently slowly with time, and its efficiency is demonstrated on some very simple examples. In order to find the laser modes, this method can be used when the mode field is not too great and does not change with time due to "burning out" of the population inversion by the induced radiation. However, it should be borne in mind that if burning out occurs, the distribution of ε usually changes more rapidly than the distribution of modes [14], and the concept of modes is no longer meaningful. Problems of this sort lie outside the scope of the theory of resonator modes, but in certain special cases calculations in the theory of lasers have been made [14, 15].

§2. The "Quasi-Classical" Approximation in the Complex Domain

Let us first consider an open resonator with plane-parallel unbounded mirrors located perpendicular to the axis of the resonator. We will see below that under the assumption of a resonator occupied by a dielectric of inhomogeneous permittivity this model can also be used to calculate the resonator modes for the case of non-plane-parallel mirrors.

Since we do not consider here effects related to polarization of the field, we take the wave field to be scalar. The wave equation for the electric field intensity E for a resonator occupied by a dielectric of permittivity ε(r) has the form

$$\frac{\varepsilon(\mathbf{r})}{c^2} \frac{\partial^2 E}{\partial t^2} = \Delta E. \tag{1}$$

The mirrors of the resonator are assumed to reflect ideally, so that the boundary conditions on the mirrors can be written as

$$E(z = 0) = E(z = l) = 0. \tag{2a}$$

Here the z axis is perpendicular to the mirrors, and l is the distance between the mirrors. The case of incomplete reflection can be reduced to the one considered by assuming that the losses upon reflection are "smeared out" inside the resonator, i.e., that the medium of the resonator is absorbing (cf. below). We are only interested in solutions which decay far away from the resonator axis, i.e.,

$$E(x = \pm \infty) = 0. \tag{2b}$$

Equation (1) combined with (2) determines the spectrum and field distribution of the resonator modes. We seek the solution in the form

$$E = e^{-i\omega t} \psi(\mathbf{r}). \tag{3}$$

Substituting (3) into (1), we obtain for ψ the Helmholtz equation

$$\Delta \psi + \frac{\varepsilon(\mathbf{r})}{c^2} \omega^2 \psi = 0. \tag{4}$$

We consider a two-dimensional problem in the z, x plane and let $\varepsilon(\mathbf{r}) = \varepsilon(x)$, i.e., the problem is homogeneous in the z direction. Writing the solution in the form

$$\psi(\mathbf{r}) = v(z)u(x) \tag{5}$$

and separating variables, we arrive at an equation for v(z):

$$\frac{\partial^2 v(z)}{\partial z^2} + k_z^2 v(z) = 0,$$ (6a)

$$v(0) = v(l) = 0.$$ (6b)

Here k_z is the wave vector in the z direction. Solving Eq. (6), we get

$$v(z) = \sin k_z z,$$ (7)

$$k_z = (2\pi/l)m,$$

where m is an integer, the so-called axial mode index.

We obtain for u(x) an equation of the time-independent Schrödinger type,

$$\frac{\partial^2 u}{\partial x^2} + k_x^2(\omega, x) u = 0$$ (8a)

with boundary condition

$$u(x = \pm \infty) = 0,$$ (8b)

where

$$k_x^2(\omega, x) = \frac{\omega^2}{c^2}\varepsilon(x) - k_z^2.$$ (9)

We remark that the one-dimensional Schrödinger equation is

$$i\hbar \frac{\partial \psi}{\partial t} = -\frac{\hbar^2}{2m}\Delta\psi + V(x)\psi,$$

where V(x) is the potential and m is the mass of the particle. Changing variables,

$$\psi = e^{-iEt/\hbar}$$

we arrive at the analogous Helmholtz equation

$$\frac{\partial^2 \varphi}{\partial x^2} + \frac{2m}{\hbar^2}(E - V(x))\varphi = 0.$$

Comparing this equation with Eq. (8), it is easy to see that the quantity $k_x^2(\omega, x)$ involving the wave vector is analogous to the quantum-mechanical quantity $2m/\hbar(E - V(x))$, and we may say that there exists an analogy between confinement of a particle in a potential well and total internal reflection (if k_x is sufficiently small or $\varepsilon' < 0$); between reflection over a well and transmission of light through an optically dense plate (interference filter); between reflection over a barrier and transmission of light through a plate of lesser optical density; and between transmission through a barrier and escape of light through a plate under conditions of total internal reflection.

The case of a complex dielectric permittivity corresponds to a complex potential in quantum mechanics. Moreover, if we seek the solution in the form $e^{-i\omega t \pm ik_x x}$, then $\varepsilon'' > 0$ corresponds to the case of wave damping ($\omega'' < 0$), and $\varepsilon'' < 0$ corresponds to wave amplification ($\omega'' > 0$). In quantum mechanics, when the potential is complex an imaginary term is added to the energy to give $E + i\Gamma$, which determines the so-called width of the level. The case $\Gamma > 0$ corresponds to decay of a quantum state whereas $\Gamma < 0$ corresponds to creation of a state.

The asymptotic WKB method, which is well known in quantum mechanics, is possible because of the existence of a large parameter in the Schrödinger equation ($\hbar \to 0$). In our case, this corresponds to $k_x^2 \to \infty$, i.e., the wavelength of the radiation satisfies $\lambda \to 0$.

In the first approximation, the WKB solution is written in the form [6-8, 10]

$$u(x) = A\frac{e^{i\int k_x dx}}{\sqrt{k_x}} + B\frac{e^{-i\int k_x dx}}{\sqrt{k_x}},$$

(10)

where the first term in (10) describes a wave traveling to the right, and the second term, a wave traveling to the left. It is immediately clear from (10) that the WKB approximation is closely related to the approximation of geometric optics. A natural criterion for the applicability of the WKB method is that the wavelength λ be small compared to the characteristic dimension l over which the potential varies (in our case, the dielectric constant ε_x):

$$l = \frac{k(x)}{\frac{\partial k(x)}{\partial x}},$$

i.e.,

$$\left|\frac{\frac{\partial k(x)}{\partial x}}{k^2(x)}\right| \ll 1.$$

(11)

Thus, the approximation is certainly inapplicable at the turning points, where $k_x(x) = 0$, i.e., $\lambda = \infty$. Correspondingly, the main difficulty in the WKB method is to obtain matching formulas with the aid of which solutions of type (10) can be pieced together from either side of a turning point.

The simplest method for piecing solutions together at a point a is to equate the solutions and their derivatives on either side of a,

$$u_1(a) = u_2(a), \qquad \frac{\partial u_1(a)}{\partial x} = \frac{\partial u_2(a)}{\partial x}.$$

However, this method is only justified when there are no turning points and is convenient, e.g., in piecing together solutions at points where the function $k_x^2(x)$ is not analytic.

Another method (the Langer method) is used to find the exact solution in a neighborhood of a turning point, and the asymptotic behavior of this solution is used to join the asymptotic expansions of type (10) on either side of the turning point.

There exists a third method due to Swann [8, 17] which in essence consists of analytically continuing the solution into the complex domain and passing around the turning points over a path in the complex plane that is far away from the turning points, where the form of the solution (10) is preserved. This process achieves joining of the solutions on either side of the turning points. Since our problem is complex from the start ($\varepsilon'' \neq 0$), we will use the last method. We give a brief exposition of it.

Assume we are given a turning point $x_\nu: k_x^2(x) = 0$ and let the root x_ν be simple, i.e., $k_x^2(x)$ can be written in the form $k_\nu^2(x) = \alpha(x - x_\nu)$ in a neighborhood of x_ν. We will pass around the point x_ν along a closed contour in the complex plane. In so doing, the phase of the solution (10)

$$\Phi = \int k_x(x)\, dx$$

will change. We construct lines on which Re $\Phi = 0$, the so-called Stokes lines (dashed lines in the figures) and the lines on which Im $\Phi = 0$, the anti-Stokes lines (solid). We draw these lines for a given concrete case with one turning point in the complex plane (Fig. 1) [8]. The wavy line denotes a cut in order to make the multivalued function (10) single valued [6-8, 17].

When the solution (10) is analytically continued along the contour L, we encounter the Stokes phenomenon [8]. The point is that on the Stokes lines [6-8, 17] one of the exponentials

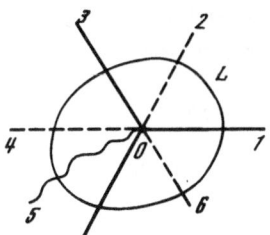

Fig. 1

decays as $|x| \to \infty$ and we must keep only the increasing solution within the accuracy of our asymptotic approximation. Thus, in the sectors on each side of a Stokes line the coefficients of the decaying exponential may be different, while they are the same for the increasing solution. For example, let

$$u(x) = \frac{A}{\sqrt{k_x}} \exp[i\Phi] + \frac{B}{\sqrt{k_x}} \exp[-i\Phi].$$

Then (cf. Fig. 1) in the sector (1, 3) the function $\exp[-i\Phi]$ increases and

$$B_3 = B_2 = B_1, \qquad A_3 = A_1 + \mu B_1,$$

where μ is an undetermined coefficient; in addition, $\exp[i\Phi]$ increases in the sector (1, 3) as $|x| \to +\infty$, and hence,

$$A_5 = A_4 = A_3, \qquad B_5 = B_3 + \lambda A_3.$$

Continuing to go around the contour L, we return to the original solution and determine λ, μ [8] to be

$$\lambda = \mu = i.$$

Since we require the solution to be bounded on line 5,

$$A_5 = 0,$$

and hence,

$$A_1 + iB_1 = 0.$$

We obtain from this a formula relating the solutions when we go from line 4 to line 1 — the solution $B_4 \exp[-i\Phi]\frac{1}{\sqrt{k}}$ (decaying as $|x| \to \infty$) on line 1 goes over into

$$\frac{A}{\sqrt{k}} (\exp[i\Phi] + i\exp[-i\Phi]) = \frac{2A}{\sqrt{k}} \exp\left[i\frac{\pi}{4}\right] \cos\left(\Phi - \frac{\pi}{4}\right) = \frac{C}{\sqrt{k}} \cos\left(\int_{x_1}^{x} k(x)\,dx - \frac{\pi}{4}\right).$$

In the case of two turning points x_1, x_2 the topology of the Stokes and anti-Stokes lines is shown in Fig. 2. In this case we obtain [8] upon going around the contour L another condition

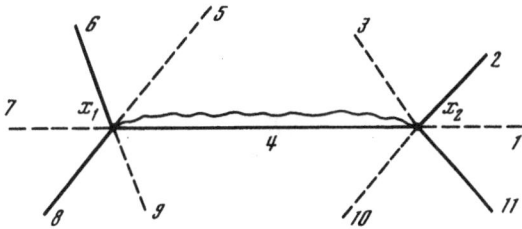

Fig. 2

for the possibility of matching a solution of type (10) decreasing at infinity on the lines 1, 7. This condition is the same as the Bohr condition,

$$\int_{x_1}^{x_2} k(\omega, x)\,dx = \pi\left(n + \frac{1}{2}\right),\tag{12}$$

where the integer n is the transverse mode index.

This condition can also be derived in a different way by using the joining formulas found previously. Thus the solution $\exp\left(i\int_{x_1}^{x} k_x dx\right)$ decaying on line 7 goes over on line 4 into the solution $\cos\left(\int_{x_1}^{x} k_x dx - \frac{\pi}{4}\right)$, and the solution $\exp\left(-i\int_{x_2}^{x} k_x dx\right)$ decaying on line 1 goes over on line 4 into the solution $\cos\left(\int_{x_2}^{x} k_x dx + \frac{\pi}{4}\right)$. But since we must obtain a single solution, it follows that the arguments of the cosines can only differ by πn, i.e.,

$$\int_{x_1}^{x_2} k_x dx - \frac{\pi}{2} = \pi n,$$

as required.

The transverse wave vector $k_x(\omega, x)$ can from (9) be written in the form

$$k_x(\omega, x) = k_z \sqrt{\frac{2\Delta\omega}{\omega_0} + \Delta\varepsilon^*(x)}.\tag{13}$$

The quantities $\omega_0 = ck_z/\sqrt{\varepsilon_0}$, $\varepsilon_0 = \varepsilon(0)$, $\Delta\omega = \omega - \omega_0$, $\Delta\varepsilon^*(x) = \varepsilon(x)/\varepsilon_0 - 1$ are in general complex.

Condition (12) determines the complex frequencies $\omega = \omega' + i\omega''$ in accordance with (9). Here ω is the real part of the mode frequency and ω'' gives the increment of the mode amplitude. In the theory of the WKB method it is proved that these conditions are applicable in calculating the mode spectrum when the distribution of the "potential" $\varepsilon(x)$ is real [6-8, 17].

If on the other hand the distribution $\varepsilon(x)$ is complex, the question of the existence of quantization conditions (12) for solutions decaying at infinity [cf. (2b)] is not trivial.

In the case of many turning points and a complicated pattern of Stokes lines, this question is unclear [13].

Indeed, for the class of solutions, e.g., decaying at infinity, the quantization conditions (12) certainly do not hold for the trivial case of reflection over a barrier in which the Stokes lines are as depicted in Fig. 3.

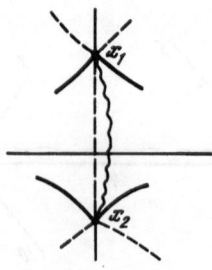

Fig. 3

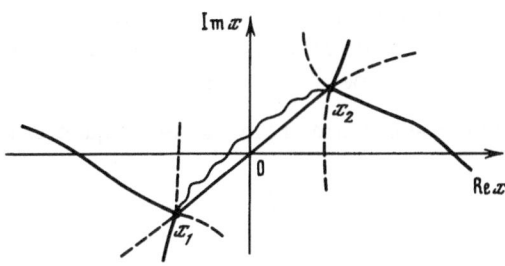

Fig. 4

Thus, the quantization condition (12) is valid in the class of solutions decaying at infinity if the topology of the Stokes levels permits. For example, in the case of two turning points, it follows from the derivation of (12) that a necessary and sufficient condition for Eq. (12) to be applicable is that the real axis lie as $|x| \to \infty$ in the sectors 2, 11 and 6, 8 (Figs. 2, 4). If this is not so, e.g., if the real axis passes through sectors 2, 3, the solution decaying on the line 1 does not become an oscillatory solution on line 4, since the joining formulas mentioned above are not applicable and we do not obtain Eq. (12).

This same question can be approached in another way, viz. by assuming that the quantization condition (12) is always satisfied. But then it is necessary to carry out an additional investigation to see for which modes (values of n) the necessary boundary conditions at infinity are satisfied.

Using the quantization formula (12) we calculate the frequency spectrum (i.e., ω' and the losses ω'') for some specific distributions $\varepsilon(x)$ commonly encountered in lasers (especially solid lasers). Calculation of the eigenmodes u(x) from (10) is in principle not complicated, but it requires a large amount of computational work using a computer, and the results of these calculations will be published later. Calculations of the mode distributions for some real $\varepsilon(x)$ were recently performed in [18].

§3. Quadratic Medium

The first typical case is that of a "lenslike," or quadratic medium, the modes of which are well known [19, 20]. This medium is often encountered in real lasers, and the quadraticity can occur either randomly [e.g., a "thermal lens" in a ruby laser [21] corresponding to a real quadratic $\varepsilon(x)$] or artificially [22]. An imaginary quadraticity can be achieved in an active medium by a quadratic transverse distribution of population inversion, which is often encountered in lasers [23, 24]. In general form, the expression for $\varepsilon(x)$ is

$$\varepsilon(x) = \varepsilon_0 + \alpha x^2 + \beta x, \tag{14}$$

where ε, ε_0 and α, β are complex.

This expression is easily rewritten as

$$\varepsilon(x) = \tilde{\varepsilon}_0 + \alpha(x - x_0)^2 = \tilde{\varepsilon}_0 + \alpha \tilde{x}^2 = \tilde{\varepsilon}_0 + \Delta\tilde{\varepsilon}(x), \tag{14a}$$

where $\tilde{\varepsilon}_0 = \tilde{\varepsilon}_0' + i\tilde{\varepsilon}_0''$, $\quad \alpha = \alpha' + i\alpha''$, $\quad x_0 = -\dfrac{\beta}{2\alpha}$,

$$\tilde{\varepsilon}_0 = \varepsilon_0 - \frac{\beta^2}{4\alpha}, \quad \tilde{x} = x - x_0. \tag{14b}$$

We will assume that $\tilde{\varepsilon}'' > 0$ (amplification).

This problem is interesting in that it permits a comparison with an exact solution (cf. [19, 20]) corresponding to a quantum-mechanical harmonic oscillator (the WKB solution for a quantum mechanical oscillator with a real potential is given in [25]). In addition, the problem of a resonator with spherical (more precisely, cylindrical) mirrors reduces to that for a resonator with a quadratic medium, as will be shown below.

We determine the frequency spectrum from the Bohr quantization conditions (12),

$$\int_{x_1}^{x_2} \sqrt{\frac{\omega^2 (\varepsilon_0 + ax^2)}{c^2} k_z^2} \, dz = \left(n + \frac{1}{2} \right) \pi. \tag{15}$$

We will take the modes to consist of paraxial waves, i.e.,

$$\frac{\omega_0}{c} \varepsilon_0' = k_z;$$

$$\left| \frac{\Delta\omega' + i\Delta\omega''}{\omega_0} \right| \ll 1, \tag{16}$$

$$\omega = \omega_0 + \Delta\omega' + i\Delta\omega''.$$

The turning points are

$$x_{1,2} = \pm \sqrt{\frac{\left(n + \frac{1}{2} \right)^2}{\frac{\omega_0}{c} \sqrt{-a}}}. \tag{17}$$

Evaluation the integral in (15) and writing out the real and imaginary parts separately in the resulting equality, we then arrive at equations for the frequency corrections $\Delta\omega'$ and $\Delta\omega''$:

$$\Delta\omega' = -\frac{\omega_0 \varepsilon_0''^2}{2 (\varepsilon_0'^2 + \varepsilon_0''^2)} + \frac{c \left(n + \frac{1}{2} \right) (u\varepsilon_0' + v\varepsilon_0'')}{\varepsilon_0'^2 + \varepsilon_0''^2}, \tag{18}$$

$$\Delta\omega'' = -\frac{\omega_0 \varepsilon_0' \varepsilon_0''}{2 (\varepsilon_0'^2 + \varepsilon_0''^2)} + \frac{c \left(n + \frac{1}{2} \right) (v\varepsilon_0' - u\varepsilon_0'')}{\varepsilon_0'^2 + \varepsilon_0''^2}, \tag{18a}$$

where

$$\sqrt{-a} = u + iv. \tag{19}$$

We consider the interesting special case when

$$\varepsilon_0'' < 0, \ \alpha' < 0, \ \alpha'' > 0,$$

and $|\alpha'| \ll |\alpha''|$.

The functions $\varepsilon'(x)$ and $\varepsilon''(x)$ are shown in Figs. 5 and 6.

From (18), (18a) we obtain

$$\Delta\omega' = \frac{c \sqrt{\alpha''} \sqrt{2} (\varepsilon_0' - \varepsilon_0'') \left(n + \frac{1}{2} \right)}{\varepsilon_0'^2 + \varepsilon_0''^2} - \frac{\omega_0 \varepsilon_0''^2}{2 (\varepsilon_0'^2 + \varepsilon_0''^2)}, \tag{18b}$$

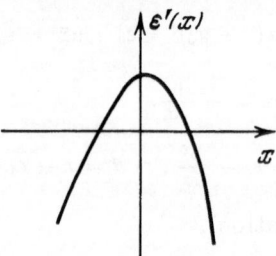

Fig. 5

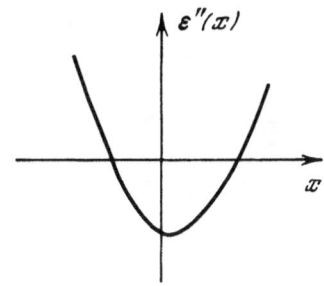

Fig. 6

$$\Delta\omega'' = -\frac{\omega_0 \varepsilon_0'' \varepsilon_0'}{2\left(\varepsilon_0'^2 + \varepsilon_0''^2\right)} - \frac{c\sqrt{\alpha''}\sqrt{2}\left(n + \frac{1}{2}\right)\left(\varepsilon_0' + \varepsilon_0''\right)}{\varepsilon_0'^2 + \varepsilon_0''^2}. \tag{18c}$$

It is thus seen from (18c) that in this case the losses increase with increasing n. The Stokes lines for this case are shown in Fig. 7. The results obtained are in fact the same as the results from exact solution of the problem [19, 20].

Since $\sqrt{-\alpha}$ and ε_0 are in general complex, so is $\Delta\omega$ and $\Delta\omega'' \neq 0$ (and depends on the transverse index n). This corresponds to the fact that modes with different transverse indices differ (are "discriminated") in terms of the losses when $v \neq 0$ or $\tilde{\varepsilon}_0'' \neq 0$. Physically, this case occurs, for example, when the transverse quadratic distribution of population inversion has a maximum on the axis of the resonator, or in a resonator with a Gaussian diaphragm whose modes are calculated in [24, 26]. In this case, the modes with the smallest losses also have the smallest excitation volume and are concentrated in the center of the inversion distribution.

On the other hand, if $\alpha'' < 0$, which corresponds to a resonator in which absorption on the axis is minimal, the losses decrease as the transverse indices and volume of excitation of modes become larger. Since under real conditions the modes with the smallest losses are excited, this will be the modes with the larger excitation volume, which are significantly influenced by the form of $\varepsilon(x)$ at larger distances from the x axis. But in real resonators, not only is $\varepsilon(x)$ far from quadratic when x is large, but diffraction effects by the boundaries of the resonator (cf. below) which are ignored in this model will play a role in mode formation. Thus, the case $\alpha'' < 0$ corresponds to a physically unrealizable situation.

As for the effect of α' on the modes, the case $\alpha' < 0$ corresponds to the model in which the modes are confined by refraction in the quadratic medium. Modes with different indices n

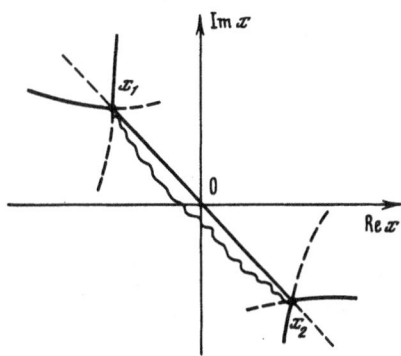

Fig. 7

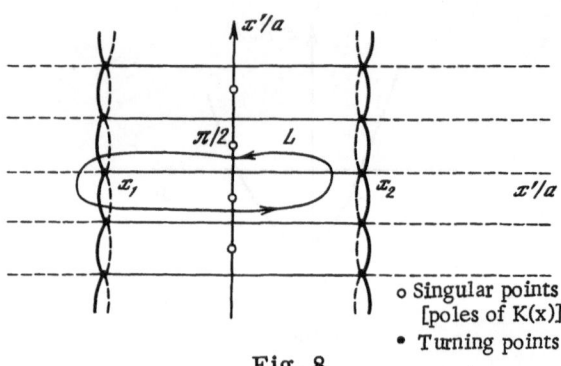

Fig. 8

may differ appreciably in the real frequencies $\Delta\omega'$. If on the other hand $\alpha' > 0$, then the $\Delta\omega'$ differ little for different n and the modes become degenerate with respect to the frequencies $\Delta\omega'$. In this case the modes are confined only by the imaginary part ε'', i.e., when $\alpha'' > 0$.

§4. Hypergeometric Layer

In this case

$$\varepsilon(x) = \varepsilon_0 + \frac{\beta}{\cosh^2 \frac{x}{a}}, \tag{20}$$

where β, a are parameters. This problem has an exact solution [3, 16]. This model gives a realistic representation of the transverse inhomogeneities of $\varepsilon''(x)$ encountered in lasers, not only at the center of the distribution but also at large distances x from the axis of the resonator. The model was used in [3, 14] in studying the effect of inhomogeneities in $\varepsilon''(x)$ on the nature of lasing in lasers.

We now show that the results obtained by the WKB method for a complex distribution differ little from the rigorous results of [3]. As in [3], let $\beta = i\delta$, $\delta > 0$. The case of a hypergeometric layer is all the more of interest to us in that there is a turning point and the pattern of the Stokes lines is quite complicated.

For example, when $\beta' > 0$, $\beta'' = 0$, the turning points are

$$\frac{x_k'}{a} = \pm \; \arccos \sqrt{\frac{-\beta'\omega_0}{\varepsilon_0\Delta\omega}}, \qquad \frac{x_k''}{a} = \pi k,$$

for $\sqrt{\dfrac{-\beta'\omega_0}{\varepsilon_0\Delta\omega}} \geqslant 1$ for the paraxial case (16) with $\Delta\omega'' = 0$, $\Delta\omega' < 0$. Here k = 0, ±1, ±2, ...; and in addition the singular points when $|k(x)| \to \infty$ are determined by the condition

$$\cosh \frac{x}{a} = 0.$$

We obtain for the latter

$$\frac{x_l'}{a} = 0, \qquad \frac{x_l''}{a} = i \frac{\pi}{a}(2l+1); \qquad l = 0, \pm 1, \pm 2, \ldots,$$

and the Stokes plane is shown in Fig. 8.

It is seen from the figure that the contour L cannot be taken to be arbitrarily far away and still go around only two turning points x_1 and x_2. For this reason, the WKB method gives approximate results even for the spectrum in this case. In fact, what follows corresponds to going around only the two turning points closest to the coordinate origin.

The integral in (12) can be evaluated in the same way as in Problem 5 in [27, p. 10]. In the paraxial case (16) we obtain the result

$$\Delta\omega_i' = - \frac{c^2}{2\omega_0\varepsilon_0 a^2}\left[\sqrt{\frac{\omega_0^2 a^2\beta}{c^2}} - \left(n+\frac{1}{2}\right)\right]^2. \tag{21}$$

Using that $\beta = -i\delta$, we have for the frequency shift $\Delta\omega'$

$$\Delta\omega_n' = - \frac{c^2}{2\omega_0\varepsilon_0 a^2}\left[\sqrt{\frac{\omega_0^2 a^2\delta}{c^2}} - \left(n+\frac{1}{2}\right)\right]^2, \tag{22}$$

and for the growth increment $\Delta\omega''$

$$\frac{\Delta\omega''\omega_0\varepsilon_0}{c^2} = \frac{1}{a^2}\sqrt{\frac{\omega_0^2 a^2\delta}{2c^2}}\left[\sqrt{\frac{\omega_0^2 a^2\delta}{2c^2}} - \left(n+\frac{1}{2}\right)\right]. \tag{23}$$

The wave vector k_x behaves asymptotically as $x \to \infty$ like

$$k_x' = \sqrt{\frac{\delta\omega_0^2}{2c^2}}\,;$$
$$k_x''(\infty) = \frac{1}{a}\left(\sqrt{\frac{\delta\omega_0^2 a^2}{2c^2}} - n - \frac{1}{2}\right). \tag{24}$$

Expressions (21)-(24) coincide with the analogous expressions in [3], provided the term $1/8$ is ignored in comparison with the dimensionless parameter of inhomogeneity

$$N^2 = \frac{1}{2}\frac{\delta\omega_0^2 a^2}{c^2}, \tag{25}$$

where, as we will show below, N is the number of modes confined by the inhomogeneity. Thus in this case the condition for the applicability of the quasiclassical approximation is

$$N \gg \frac{1}{2\sqrt{2}}. \tag{26}$$

Since at $+\infty$ our solution has the form $e^{-i\omega t + ik_x(\infty)x}$, for $k_x''(\infty) > 0$ the solution will decay in space as $x \to +\infty$, and when $k_x''(\infty) < 0$, it will increase. Analogously, when $\Delta\omega'' > 0$ the solution will decay in time, while when $\Delta\omega'' < 0$ it will increase

We remark that it follows from (23) and (24) that

$$\text{sgn}\,\Delta\omega'' = \text{sgn}\,k_x''(\infty). \tag{27}$$

In other words, a solution finite at $+\infty$ decays in time, where it increases with time if it is unbounded at $+\infty$. It follows at once from (26) that the bounded solutions as in [3] satisfy the condition

$$\left(n+\frac{1}{2}\right)^2 \leqslant \frac{\delta\omega_0^2 a^2}{2c^2} = N^2, \tag{28}$$

i.e., N does indeed determine the number of modes confined by the inhomogeneity. We remark that the quasiclassical condition is simply that $n \gg 1$. In this case quasiclassical arguments are applicable, and for $n = 0$ it is important only that $N \gg 1/(2\sqrt{2})$. When the parameter N increases, so does the number of modes, and as follows from (25), the mode losses increase with n. It also follows from (25) that, as N gets larger, a mode of given index n depends more strongly on the inhomogeneity. All this coincides with the precise results of [3] when

$N \gg 1$. A discernible discrepancy can occur only at small N, e.g., when only the single mode with n = 0 is confined.

§5.   Resonator with Nonplanar Mirrors and Lenses

The theory discussed above was developed for a resonator with plane-parallel unbounded mirrors occupied by a dielectric with permittivity ε and transverse inhomogeneities ("the distributed model"). Such a model can also be used to calculate the modes of resonators with local components (flat mirrors, lenses, diaphragms) along the axis, under the assumption that the transformation of the phase and amplitude of a wave after a double passage across the resonator in the distributed model is the same as the transformation of a wave in a resonator with local components. It was proved in [28] that a resonator with nonplanar mirrors and a resonator with an inhomogeneous medium are equivalent in the two-dimensional case. In the equivalent distributed model, the phase and amplitude inhomogeneities of a local resonator are "smeared out" along the axis of the resonator. It is obvious that the field distribution in the two models will not differ significantly, provided the additional phase change due to variation in the transverse wave vector k_x during the transverse displacement occurring during a double passage of the wave between the mirrors is much less than π. For example, if a local component of the resonator causes a sufficiently "smooth" complex transformation of the wave whose distribution in a transverse section is

$$\rho\,(x) = e^{i\sigma'(x) - \sigma''(x)} = e^{i\sigma(x)},$$

the additional shift of the wave after a double passage across the resonator, proportional to the turning angle of the wave front at the local inhomogeneity, is

$$\frac{L}{k}\left|\frac{d\sigma\,(x)}{dx}\right|,$$

and for the distributed model to be applicable, it is necessary that

$$\frac{L}{k}\left|\frac{d\sigma\,(x)}{dx}\right|k_x \ll \pi \tag{29}$$

(L is the length of the resonator). An inhomogeneous (planar) wave transformer is approximated here by a distribution*

$$\varepsilon\,(x) = 1 + \Delta\varepsilon\,(x),$$

where

$$\Delta\varepsilon\,(x) = \frac{\sigma\,(x)}{kL} = \frac{\sigma'}{kL} + \frac{i\sigma''}{kL}. \tag{30}$$

Condition (28) can be written as

$$\left|\frac{d\Delta\varepsilon\,(x)}{dx}\right| \ll \left|\frac{\pi}{k_x L^2}\right|. \tag{29a}$$

It can be shown analogously that when a local inhomogeneity is of an irregular character, it cannot be allowed for in the distributed model if its dimension d is very small, i.e.,

$$d < \left|\frac{2Lk_x}{k}\right|. \tag{31}$$

---

*In particular, if the amplitude coefficient of reflection r of the mirror is not equal to unity in absolute value, then $\sigma'' = \ln|r|$ and $\Delta\varepsilon'' = \ln|r|/kL$.

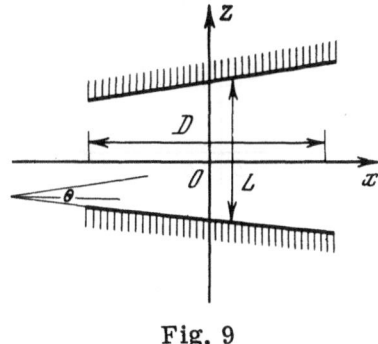

Fig. 9

Examples of some equivalent resonators are given in Figs. 9, 10, and 11.

As an example we consider a resonator formed by cylindrical mirrors of the same radius R located a distance L apart, which is equivalent to a resonator with a quadratic medium (cf. Fig. 10) in which

$$\varepsilon\,(x) = 1 + \alpha x^2, \tag{32}$$

if we put

$$\alpha = -\frac{1}{2RL}\,. \tag{33}$$

Moreover, $\alpha < 0$ if the mirrors are concave and $\alpha > 0$ if they are convex. Analogously, a resonator with a lens of focal length F is also equivalent to a resonator with a quadratic medium, and

$$\alpha = -\frac{1}{LF}\,. \tag{34}$$

In addition, it is shown in [26, 24] that a diaphragm with a Gaussian transverse distribution of transmission

$$\eta = \eta_0 e^{-x^2/b^2},$$

where b is the effective radius of the diaphragm, can be treated as a lens with an imaginary focal length

$$F = ikb^2.$$

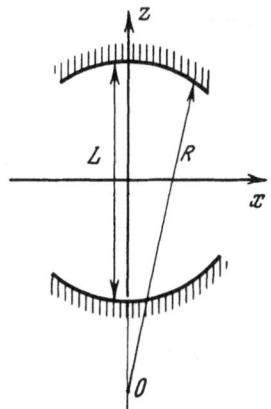

Fig. 10

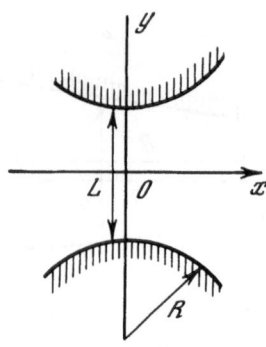

Fig. 11

Then for the corresponding equivalent resonator with quadratic medium,

$$a = ia'' = \frac{i}{kb^2L}.$$

(34a)

Analogously, a resonator with mirrors with reflection coefficient depending on the transverse coordinate according to a Gaussian law, the modes of which are calculated in [29], is also equivalent to a resonator with a complex quadratic medium.

The use of the distributed model with a quadratic medium for calculating the modes of resonators with cylindrical mirrors and lenses is restricted by the condition (29), which says that in this model only modes for which the transverse dimension $2a$ is not very large can be calculated. We obtain from (29) the condition

$$a < \sqrt[4]{\frac{\pi^2}{k^2 L^4 \alpha^3}}$$

(35)

for the largest transverse dimension $2a$ of the modes, or (for a given dimension a) a condition on the minimal focal length F of the lens,

$$F > \sqrt{\frac{4La^4}{\lambda^2}}.$$

(35a)

For example, at $\lambda = 693$ nm (ruby laser), $L = 40$ cm, $a = 0.2$ cm, F must be greater than 400 cm. When these conditions hold, the resonator modes and their spectrum can be calculated using the distributed model according to the formulas in [13–17].

§6. Mirrors and Diaphragms with Sharp Boundaries

The WKB method makes no allowance for diffraction effects occurring in resonators with bounded mirrors or with internal diaphragms with sharp boundaries. In the presence of diffraction, radiation is ejected into the interior of the resonator, a process described in [1, 30] as reflection and transformation of paraxial waves from the sharp boundaries and diaphragms. It is shown there that at small angles between the wave and resonator axis (more precisely, under the condition $|k_x| \ll \sqrt{k/L}$) reflection of waves occurs mainly on the boundaries of the mirrors, and the coefficients of transformation of waves into waves with other axial indices (and conversely) is much less than the coefficient of reflection, which is equal to

$$r_{00} = e^{-k_x 0.824 \sqrt{\frac{L}{k}}(1+i)}$$

(36)

for a wave with transverse wave vector k_x. This result makes it possible to reduce the problem of a resonator with mirrors having sharp boundaries to the case of a resonator with unbounded mirrors filled with a laminar complex dielectric, with a sharp boundary between the

layers parallel to the axis of the resonator. A similar procedure was used in [31] in finding
the modes in the special case of a resonator with inclined mirrors. Taking the jump $\Delta\varepsilon$ on
this boundary to give a Fresnel coefficient of reflection equal to r_{00} in accordance with [36]
(as $k_x \to 0$), we obtain the value

$$\Delta\varepsilon_c = -\frac{2i}{(0.824)^2 kL} = -i\frac{2.946}{kL} \tag{37}$$

for the jump. Thus, we must calculate the modes for a "distributed" model for a resonator
containing a piecewise continuous medium with a complex dielectric permittivity. The solutions in each of the regions of analyticity can be found by the WKB method, and they are joined
at the points of discontinuity by equating the solutions and their derivatives on either side of
a point of discontinuity. We remark that in this case the turning points (caustics) may not
exist and the mode may be confined by the jump $\Delta\varepsilon$ (by the boundaries of the mirror). Piecing
the solutions together in the case of a piecewise continuous $\Delta\varepsilon$ using (15), we obtain in place
of (12) the generalized quantization conditions

$$\int_{x_1}^{x_2} k_x(\omega, x)\, dx = (n + \tilde{n}(x_1) + \tilde{n}(x_2))\,\pi, \tag{38}$$

where

$$\tilde{n}(x_i) = \begin{cases} \frac{1}{4} \text{ for an ordinary linear turning point;} \\ \dfrac{1}{\pi}\arctan i\dfrac{k_{2x}(\omega, x_i)}{k_{1x}(\omega, x_i)} \text{ if } x_i \text{ corresponds to the boundary of the mirror;} \end{cases}$$

$$k_{1x}(\omega, x_i) = k_z\sqrt{\frac{2\Delta\omega}{\omega_0} + \Delta\varepsilon(x_i)},$$

$$k_{2x}(\omega, x_i) = k_z\sqrt{\frac{2\Delta\omega}{\omega_0} + \Delta\varepsilon(x_i) + \Delta\varepsilon_c}, \tag{39}$$

where $\Delta\varepsilon(x_i)$ is found from (13), and $\Delta\varepsilon_c$ from (37). The values of k_{1x} and k_{2x} are the values
of k_x at the point x_i as $x - x_i \to \pm 0$.

§7. Resonator with Inclined Mirrors and Linear

Transverse Variation of the Absorption

A resonator with flat mirrors making an angle θ with each other (cf. Fig. 9) is approximated by a distributed model in which the permittivity depends linearly on the transverse coordinate x (the coordinate origin is at the center of the mirrors which have dimension D),

$$\varepsilon = 1 + \beta x, \tag{40}$$

where

$$\beta = 2\theta/L.$$

The boundaries of the mirrors correspond to the jump $\Delta\varepsilon$ as given by (37). The case
where the absorption coefficient changes linearly with x corresponds to an imaginary coefficient
$\beta = i\beta''$, so that we will in the general case consider β to be complex, $\beta = \beta' + i\beta''$.

The field of a mode formed by the boundaries of the mirrors in any case lies in the direction in which the mirrors open out. On the other hand, a caustic forms (cf. Fig. 9) for "caustic" modes if the coordinate of a turning point satisfies

$$\operatorname{Re} x_2 < \frac{D}{2} \tag{41}$$

(here D/2 is the coordinate of the boundary of the mirror). When this condition does not hold,
the mode will be confined by the boundary of the mirror ("noncaustic" mode) similarly to what

happens in a resonator with plane-parallel mirrors. Using (9), (15) we find from the equation

$$k_x^2(x_2) = 0$$

that

$$x_2 = -\frac{\Delta\omega_n}{\omega_0\beta}. \tag{42}$$

Thus, the position of the caustic depends on the mode number n. In the case of "caustic" modes, we must put

$$x_1 = \frac{D}{2}, \qquad x_2 = -\frac{\Delta\omega_n}{\omega_0\beta}, \qquad \tilde{n}(x_2) = \frac{1}{4} \tag{43}$$

in (38), and when $|\Delta\varepsilon_c| \gg \left|\frac{2\Delta\omega_n}{\omega_0}\right|$,

$$\tilde{n}(x_1) = -\frac{1}{\pi}\arctan\sqrt{\frac{\omega_0\Delta\varepsilon_c}{2\Delta\omega_n}} = \frac{1}{2} - \frac{1}{\pi}\arcsin\sqrt{\frac{2\Delta\omega_n}{\omega_0\Delta\varepsilon_c}} \approx \frac{1}{2} - \frac{1}{\pi}\sqrt{\frac{2\Delta\omega_n}{\omega_0\Delta\varepsilon_c}}. \tag{43a}$$

For "noncaustic" modes,

$$\tilde{n}(x_1) + \tilde{n}(x_2) = 1 - \frac{2}{\pi}\arcsin\sqrt{\frac{2\Delta\omega_n}{\omega_0\Delta\varepsilon_c}} \approx 1 - \frac{2}{\pi}\sqrt{\frac{2\Delta\omega_n}{\omega_0\Delta\varepsilon_c}}. \tag{44}$$

Using (43) we obtain from Eq. (38). a formula for the mode frequencies (approximate when $\left|\Delta\varepsilon_c\right| > \left|\frac{2\Delta\omega_n}{\omega_0}\right|$)

$$\Delta\omega_n = \frac{\omega}{2}\left[\frac{3\pi(-\beta)}{2k}\left(n + \frac{3}{4}\right)\right]^{2/3} - \frac{c\beta}{2}\frac{1}{\sqrt{\Delta\varepsilon_c}} - \frac{\beta D\omega_0}{4} \tag{45}$$

for the "caustic" modes, and the distance of the caustic from the boundary of the mirrors is (approximately)

$$d_n = \mathrm{Re}\left[\frac{1}{(-\beta)^{1/3}}\left(\frac{3\pi}{2k}\left(n + \frac{3}{4}\right)\right)^{2/3}\right]. \tag{46}$$

The condition for "caustic" modes is $d_n \ll D$. The losses of such modes are given by

$$\Delta\omega'' = \frac{\omega_0}{2}\left[\frac{3\pi}{2k}\left(n + \frac{3}{4}\right)\right]^{2/3}(\beta'^2 + \beta''^2)^{1/3}\sin\left(\frac{2}{3}\arctan\frac{\beta''}{\beta'}\right) - 0.206c(\beta' - \beta'')\sqrt{k}L - \frac{\beta'' D\omega_0}{4}. \tag{47}$$

If $\beta'' = 0$ (i.e., there is no absorption), then

$$\Delta\omega'' = -0.412c\theta\sqrt{\frac{k}{L}}, \tag{47a}$$

i.e., the losses do not depend on the mode number and are proportional to the angle of inclination of the mirrors. This result and Eq. (47a) agree with the results of [32]. When $\beta'' \neq 0$, the degeneracy of the modes with respect to losses is lifted in accordance with (47), and if the absorption decreases toward the boundary of the mirror the mode losses increase with n, whereas the losses decrease in the opposite case.

For "noncaustic" modes of form similar to the modes of a plane-parallel resonator, when $d_n > D$, i.e., when

$$\left(n + \frac{3}{4}\right) > \frac{2kD}{3\pi}\sqrt{|\beta|D},$$

we obtain for $\Delta\omega_n$ using (38) and (44) the approximate expression

$$\Delta\omega_n = \frac{\pi^2 (n+1)^2 c}{2kD^2} \left[1 - \frac{4}{Dk} \frac{1}{\sqrt{\Delta\varepsilon_c}} \left(1 + \frac{\beta^2 k^4 D^6}{24\pi^4 (n+1)^4} \right) \right]. \tag{48}$$

Hence the mode losses for $\beta'' = 0$ are

$$\omega'' = -\frac{\pi^2 (n+1)^2 c^3}{\omega^2 D^3} 0.824 \sqrt{kL} \left(1 + \frac{\theta^2 k^4 D^6}{6\pi^4 (n+1)^4 L^2} \right). \tag{49}$$

This result is close to the result of [33]. The small difference in the numerical coefficient in the second term in brackets clearly arises because the calculations in [33] are for round, not flat, mirrors. It is easy to see that as $\beta \to 0$ ($\theta \to 0$), Eqs. (48), (49) go over into the equations for a plane-parallel resonator given, e.g., in Vainshtein [1].

It should be added that the distributed model is suitable for calculating the modes of a resonator with inclined mirrors when conditions (29) are satisfied, which in the present case have the form

$$\theta < \frac{1}{\sqrt[4]{3\left(n+\frac{3}{4}\right)}} \sqrt{\frac{\lambda}{2L}} = \frac{1}{2} \sqrt[3]{\frac{\lambda^2}{Ld_n}} \tag{50}$$

(for "caustic" modes with $d_n < D$). An analogous condition can also be derived for "non-caustic" modes. It follows, e.g., that for $\lambda = 693$ nm the theory is applicable for angles θ smaller than 1.5' (for $d_n = 0.8$ mm). In addition, it should be remarked that even when the model is applicable, the condition

$$|k_x| < \sqrt{\frac{k}{L}}, \tag{51}$$

necessary for the applicability of Eq. (36), fails for large transverse indices n. For this reason, when

$$\left(m + \frac{3}{4}\right) > \frac{1}{3\pi\theta \sqrt{RL}}, \tag{51a}$$

Eqs. (45) and (48) also become approximate for "caustic" modes. These restrictions coincide with the restrictions in deriving the equations for a plane-parallel resonator in [1, 30].

§8. General Case, Small Imaginary Part

We have seen from the preceding examples by comparison with precise results obtained by more general approximation methods that the WKB method gives a good approximation for the mode frequencies and losses. In the general case, the evaluation of the integral in (12) is quite complicated. However, if the imaginary part $i\Delta\varepsilon''(x)$ is small enough and the turning points x_1 and x_2 lie sufficiently close to the real axis, Eq. (12) is easily seen to split into the equation

$$\int_{\mathrm{Re}\, x_1}^{\mathrm{Re}\, x_2} dx \sqrt{\mathrm{Re}\, k_x^2 (\omega_0 + \Delta\omega', x)} = \pi\left(n + \frac{1}{2}\right). \tag{52}$$

which determines $\Delta\omega'$, and the equation

$$\Delta\omega'' = \int_{\mathrm{Re}\, x_1}^{\mathrm{Re}\, x_2} dx \frac{\mathrm{Im}\, k_x^2 (\omega_0 + \Delta\omega', x)}{\sqrt{\mathrm{Re}\, k_x^2 (\omega_0 + \Delta\omega', x)}} \left\{ \int_{\mathrm{Re}\, x_1}^{\mathrm{Re}\, x_2} dx \frac{1}{\sqrt{\mathrm{Re}\, k_x^2 (\omega_0 + \Delta\omega', x)}} \frac{\partial\, \mathrm{Re}\, k_x^2 (\omega_0)}{\partial x} \right\}^{-1}, \tag{53}$$

which determines $\Delta\omega''$.

§9. Calculation of the Modes of a Resonator with a Quadratic Medium

Calculation of the eigenfunctions of a resonator by the WKB method is computationally quite complicated and is possible in practice only with the aid of a computer. We give an example of such a calculation and compare it with precise results in the case of a quadratic medium. The resonator modes for a quadratic medium are well known — they are the Hermite polynomials (the wave functions of a harmonic oscillator) [16]. We constructed graphs of the exact solutions using a computer as well as the WKB approximations for the case of both real and complex ε. The WKB solutions were normalized so as to make the WKB amplitudes of the central field maximum (or the derivative of the amplitude at the center of the resonator) coincide with the corresponding values in the exact solution. Figure 12a-d shows the amplitudes y of the exact solution (curve 1) and the WKB approximation (curve 2) to arbitrary scale for mode indices n = 0, 1, 2, and 4 respectively for the case of real ε; $x_{1,2}$ are the turning points. Figure 13 shows the mode for n = 8. In this case it is no longer assumed that ε is real and the amplitude $y = Ae^{i\varphi}$ is in general complex. Therefore, the graphs simultaneously show the modulus squared $|y|^2$ of the amplitude (curve 1 for the exact solution, curve 2 for the WKB solution) and phase Φ (curves 3 and 4 for the exact and WKB solutions, respectively); $\tilde{x}_{1,2}$ are the points of intersection of the Stokes lines with the real axis. Figure 13a corresponds to the case in which Im ε = 0; Fig. 13b to $|\operatorname{Im}\alpha| \ll |\operatorname{Re}\alpha|$ (cf. (14b); in Fig. 13c, $\alpha = e^{i\varphi}$, $\beta = e^{i\psi}$, $\varphi = \psi = 0.1\pi$ [cf. (14a), (14b)]; in Fig. 13d, $\alpha = i$. Figure 14 shows the case $\alpha = i$ for n = 0. The behavior of the modes in the case of complex ε corresponds to the results of Sec. 3. The graphical examples indicate that the approximate WKB solutions show an extremely satisfactory agreement with the exact solutions for real and complex ε even for small n, and the agreement improves as n increases.

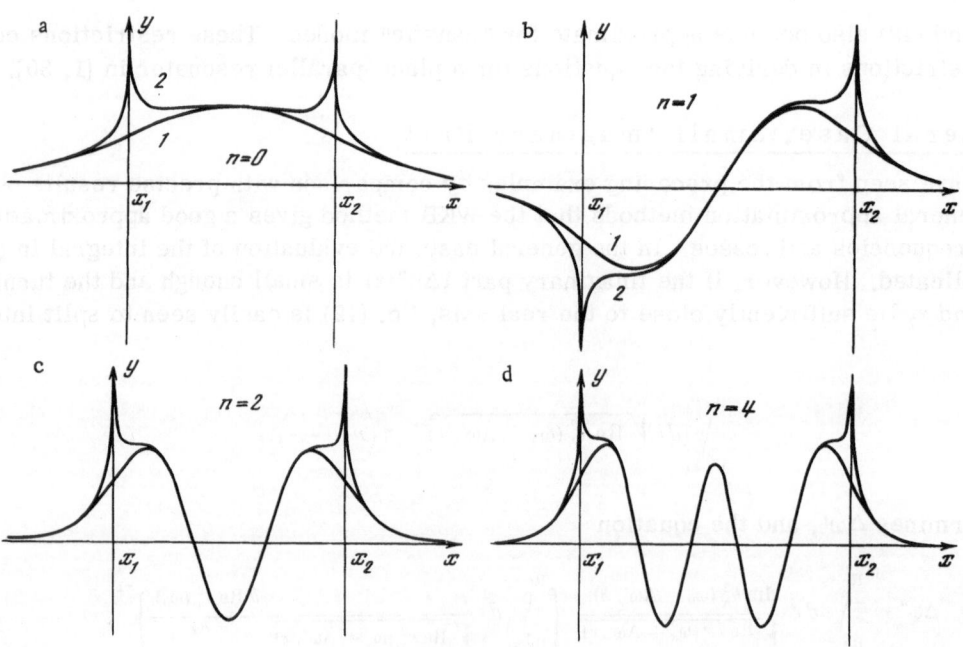

Fig. 12

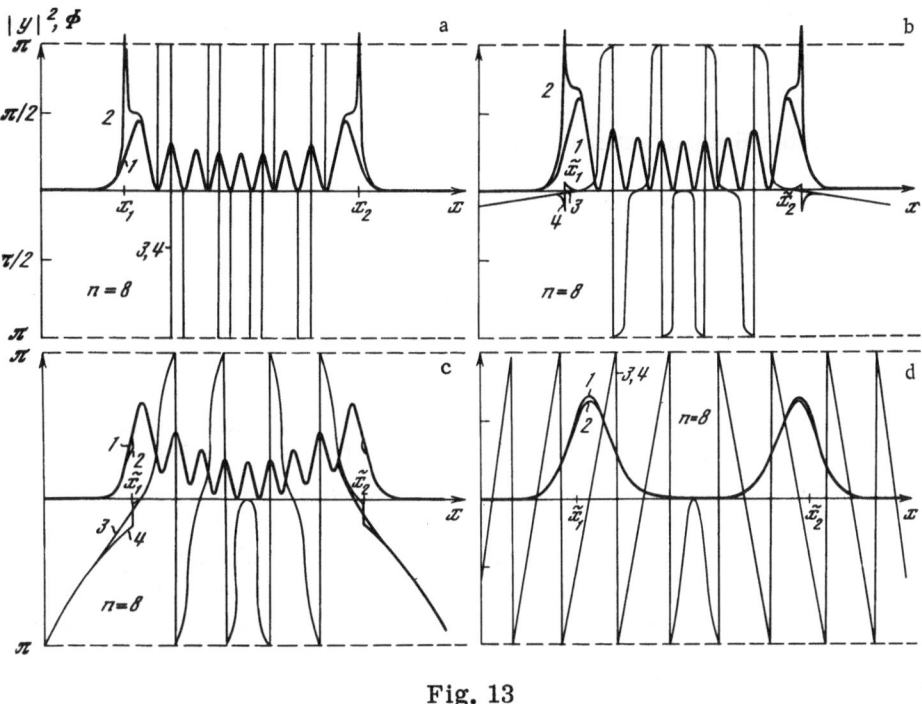

Fig. 13

§10. Discussion of Results. Limits of Applicability
of the Theory

It follows from the results obtained above that the form, frequency spectrum, and mode losses calculated by the WKB method for different resonators in the complex case show good agreement with the values obtained by exact or more general approximate methods [1, 32, 33]. This indicates that the WKB method can be used efficiently to calculate the modes of various oscillators with an accuracy which is completely satisfactory for experimental applications.

Let us discuss the accuracy of the WKB method in more detail. It is known [34] that the accuracy of the WKB method is determined by the size of the terms discarded in the expression for the wavefunction (10). This expression has the order of magnitude $1/|k_x \Delta x|^2$, where Δx is

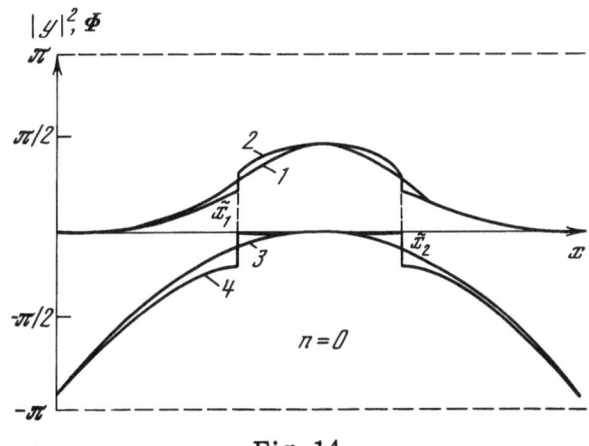

Fig. 14

the distance between turning points. Thus, the condition for the WKB method to be applicable is

$$\frac{1}{|k_x \Delta x|^2} \ll 1.$$

(54)

It follows from the Bohr quantization conditions that

$$\left| \int_{x_1}^{x_2} k_x \, dx \right| \sim |k_x \Delta x| \sim \left(n + \frac{1}{2} \right) \pi,$$

i.e., the criterion (54) holds best for large mode indices n. We·remark that the accuracy as determined by (54) is also quite good for small n: when n = 0, $|k_x \Delta x|^{-2} \approx 0.4$; when n = 1, $|k_x \Delta x|^{-2} \approx 0.04$; etc.

It has been remarked repeatedly in the literature [6, 7, 17] that the accuracy of the WKB method is better than is given by the estimate of type (54). The reason is connected with the asymptotic character of the series expansion of the solution in the WKB method. The quantization conditions (12) hold with great accuracy in those cases where, in deriving the quantization conditions by the Swann method, we can pass around a contour enclosing the turning points and sufficiently far from them in the complex x plane (the WKB solution tends asymptotically to the exact solution). For example, for the case of a quadratic medium (a harmonic oscillator) the conditions are in general exact, since there are no other turning points in the complex plane and the contour traversed around the only pair of turning points can be chosen arbitrarily far away, although the turning points themselves may be very close. In addition, if the "potential" well $k_x^2(x)$ is sufficiently symmetric and mildly sloping without sharp "sides," the largest contribution to the integral in (12) comes from the region near the bottom of the well, while the regions near the turning points give small contributions [17]. Hence the quantization condition (12) can again hold with great accuracy. We saw a·specific instance of this in the example of a hypergeometric layer, where the accuracy of the WKB method was better than 1/n. All these results give a basis for the supposition that in cases of interest for practical applications, the WKB method usually possesses a greater accuracy than corresponds to estimates of the type (54).

We have seen that the one-dimensional WKB method with a complex distribution of the dielectric permittivity can be used to calculate the modes of two-dimensional resonators with bounded nonplanar mirrors, lenses, and diaphragms, etc., by reducing the latter problems to an equivalent "distributed" model, in which the local phase and amplitude variations are "smeared out" inside the resonator and the mirror boundaries are replaced by a step-jump of the permittivity (37). For this to be possible, condition (29) must hold. As for condition (51), which is usually more stringent, it is only required in order to make use of equations of types (45), (48), and is not necessary for the applicability of the distributed model itself.

It should be noted that the distributed model is the simplest approximation of a multidimensional resonator with regular distribution of ε (x, y, z) and/or nonplanar mirrors (e.g., spherical mirrors of arbitrary radius of curvature), and the problem can be reduced to several one-dimensional problems if it is possible to introduce corresponding curvilinear coordinates and use the method of separation of variables [16]. Such a method was used to calculate an empty resonator with ellipsoidal mirrors in [10]. Thus, the complex WKB method can in general be used for quite accurate calculations of various multidimensional resonators containing an active medium of regular type.

It is shown in [35, 36] that the effect of an inhomogeneous dielectric permittivity of irregular, random character, such as occurs in the media of lasers, is, first, to cause the appearance of additional scattering losses (the same for all modes), and second, to cause the appearance of spreading out and frequency shifts of individual modes. In the framework of a model with a distributed medium, the appearance of additional losses can be described by the addition to ε of a constant imaginary term. Under condition (31), each inhomogeneity of this type with a small dimension d gives an average contribution to ε of the order

$$i \frac{2.046}{kL} (1 - e^{-\sigma^2}) \frac{d}{D},$$

where σ is the average phase shift or amplitude change due to the inhomogeneity; D is the transverse diameter of the resonator. Moreover, the distribution $\varepsilon(x)$ must be smoothed by averaging over steps of length determined by (31). It should be remarked that the averaging step depends on k_x, i.e., also on the transverse mode index.

Thus the WKB method is extremely universal, simple, and accurate in calculating the modes for a large class of optical resonators with active media. Another valuable aspect of the method is the simplicity and transparency of the results in terms of the light ray representation of a mode field.

The authors express their thanks to L. A. Vainshtein, Corresponding Member of the Academy of Sciences of the USSR, for a fruitful discussion of this work.

LITERATURE CITED

1. L. A. Vainshtein, Open Resonators and Open Waveguides [in Russian], Sov. Radio (1966).
2. L. Ronchi, Optical Resonators, in: Laser Handbook, F. T. Arecehi and E. O. Schultz-Budois, eds., North-Holland, Amsterdam (1972), p. 153.
3. T. I. Kuznetsova, Zh. Éksp. Teor. Fiz., 34:419 (1964); Zh. Éksp. Teor. Fiz., 36:62; Trudy FIAN, 43:116 (1968).
4. Yu. A. Anan'ev, V. V. Lyubimov, and I. B. Orlova, Zh. Éksp. Teor. Fiz., 39:1872 (1962).
5. Yu. A. Kalinin, A. A. Mak, and A. I. Stepanov, Zh. Tekh. Fiz., 38:1108 (1968).
6. J. Heading, An Introduction to Phase-Integral Methods, Wiley, New York (1962).
7. I. Freeman and P. U. Freeman, The WKB Approximation [Russian translation], Mir, Moscow (1967).
8. G. M. Zaslavskii, Lectures on the Application of the WKB Method in Physics [in Russian], Novosibirsk (1965).
9. J. B. Keller and S. I. Rubinow, Ann. Phys., 9, 1:24 (1960).
10. V. M. Bykov and L. A. Vainshtein, Zh. Éksp. Teor. Fiz., 20:338 (1965).
11. V. P. Silin, FIAN Preprint No. 211, Moscow (1962).
12. M. V. Fedoryuk, Appendix to Asymptotic Expansions of Solutions of Ordinary Differential Equations by W. Wasow [Russian translation], Mir, Moscow (1968).
13. Yu. N. Dnestrovskii and D. P. Kostomarov, Zh. Vychisl. Mat. Mat. Fiz., 4:267 (1964).
14. V. S. Letokhov and A. F. Suchkov, Zh. Éksp. Teor. Fiz., 50:1148 (1965); Zh. Éksp. Teor. Fiz., 52:282 (1967); Trudy FIAN, 43:167 (1968).
15. S. I. Vlasov, V. I. Talanov, and A. I. Khizhnyak, Izv. Vyssh. Uchebn. Zaved., Radiofiz., 14:570 (1971); 16:828 (1971).
16. L. D. Landau and E. M. Lifshitz, Quantum Mechanics, Nonrelativistic Theory, Pergamon Press, Oxford—New York (1977).
17. W. H. Farry, Phys. Rev., 71:360 (1947).
18. A. Dedoen, Opt. Commun., 12:329 (1974).
19. H. Kogelnik, Appl. Opt., 4:1562 (1965); H. Kogelnik and T. Li, Appl. Opt., 5:1550 (1966).
20. D. Marcuse, Light Transmission Optics, Chap. 7, Van Nostrand Reinhold, New York (1972).
21. A. P. Veduta, A. M. Leontovich, and V. N. Smorchkov, Zh. Éksp., Teor. Fiz., 48:87 (1965).

22. I. Kitano, K. Koizurai, K. H. Matsumuar, et al., J. Jpn. Soc. Appl. Phys. Suppl., 39:63 (1970).
23. G. Lampis, C. A. Sacchi, and O. Svelto, Appl. Opt., 3:1467 (1964); N. G. Bondarenko and I. V. Eremina, Zh. Prikl. Spektrosk., 8:599 (1968); L. W. Casperson and A. Yariv, Appl. Phys. Lett., 12:355 (1968).
24. A. M. Leontovich and V. L. Churkin, Zh. Éksp. Teor. Fiz., 59:7 (1970).
25. S. Flügge, Practical Quantum Mechanics, Springer-Verlag, New York (1971).
26. V. I. Talanov, Izv. Vyssh. Uchebn. Zaved., Radiofiz., 8:260 (1965); V. S. Averbakh, S. N. Vlasov, and V. I. Talanov, Zh. Éksp. Teor. Fiz., 36:497 (1966).
27. P. I. Gol'dman and V. D. Krivchenkov, Problems in Quantum Mechanics [in Russian], Fizmatgiz, Moscow (1966).
28. B. L. Rozhdestvenskii, Zh. Tekh. Fiz., 23:1609 (1953).
29. S. N. Vlasov and V. I. Talanov, Radiotekh. Élektron., 10:55 (1965); S. N. Vlasov, Radiotekh. Élektron., 10:715 (1965); N. G. Vakhitov, Radiotekh. Élektron., 10:1676 (1965).
30. L. A. Vainshtein, Zh. Éksp. Teor. Fiz., 44:1056 (1963); Zh. Tekh. Fiz., 34:193 (1964).
31. A. F. Suchkov, Trudy FIAN, 43:161 (1968).
32. V. V. Lyubimov and I. B. Orlova, Zh. Tekh. Fiz., 39:2183 (1969); Opt. Spektrosk., 30:758 (1971).
33. H. Ogura, Y. Yoshida, and J. Ikena, J. Phys. Soc. Jpn., 20:598 (1965).
34. A. B. Migdal and V. P. Krainov, Approximate Methods in Quantum Mechanics, Chap. 3 [in Russian], Nauka, Moscow (1966).
35. V. V. Lyubimov, Opt. Spektrosk., 21:22 (1966); Yu. A. Anan'ev, V. V. Lyubomov, and B. M. Sedov, Zh. Prikl. Spektrosk., 8:955 (1968).
36. S. M. Kozel and G. R. Lokshin, Radiotekh. Élektron., 19:1142 (1974).